Revealing the Invisible Mine

Pacific Perspectives

Studies of the European Society for Oceanists

General Editors:

Edvard Hviding, University of Bergen
Toon van Meijl, Radboud University

Oceania is of enduring contemporary significance in global trajectories of history, politics, economy and ecology and has remained influential for diverse approaches to studying and understanding human life worlds. The books published in this series explore Oceanic values and imaginations, documenting the unique position of the Pacific region – its cultural and linguistic diversity, its ecological and geographical distinctness and always fascinating experiments with social formations. This series thus conveys the political, economic and moral alternatives that Oceania offers the contemporary world.

Volume 8
Revealing the Invisible Mine
Social Complexities of an Undeveloped Mining Project
Emilia E. Skrzypek

Volume 7
If Everyone Returned, the Island Would Sink
Urbanisation and Migration in Vanuatu
Kirstie Petrou

Volume 6
Pacific Realities
Changing Perspectives on Resilience and Resistance
Edited by Laurent Dousset and Mélissa Nayral

Volume 5
In the Absence of the Gift
New Forms of Value and Personhood in a Papua New Guinea Community
Anders Emil Rasmussen

Volume 4
Living Kinship in the Pacific
Edited by Christina Toren and Simonne Pauwels

Volume 3
Belonging in Oceania
Movement, Place-Making and Multiple Identifications
Edited by Elfriede Hermann, Wolfgang Kempf and Toon van Meijl

Volume 2
Pacific Futures
Projects, Politics and Interests
Edited by Will Rollason

Volume 1
The Ethnographic Experiment
A.M. Hocart and W.H.R. Rivers in Island Melanesia, 1908
Edited by Edvard Hviding and Cato Berg

Revealing the Invisible Mine

Social Complexities of an Undeveloped Mining Project

Emilia E. Skrzypek

berghahn
NEW YORK • OXFORD
www.berghahnbooks.com

First published in 2021 by
Berghahn Books
www.berghahnbooks.com

Library of Congress Cataloging-in-Publication Data

Names: Skrzypek, Emilia E., author.

Title: Revealing the invisible mine : social complexities of an undeveloped mining project / Emilia E. Skrzypek.

Description: New York : Berghahn Books, 2021. | Series: Pacific perspectives : studies of the European Society for Oceanists ; volume 8 | Includes bibliographical references and index.

Identifiers: LCCN 2020016102 (print) | LCCN 2020016103 (ebook) | ISBN 9781789208566 (hardback) | ISBN 9781789208573 (ebook)

Subjects: LCSH: Frieda River Copper-Gold Project (Papua New Guinea)—Social aspects. | Mineral industries—Social aspects—Papua New Guinea—East Sepik Province. | Mines and mineral resources—Social aspects—Papua New Guinea—East Sepik Province. | Papua New Guineans—Papua New Guinea—East Sepik Province. | East Sepik Province (Papua New Guinea)—Social conditions. | East Sepik Province (Papua New Guinea)—Economic conditions.

Classification: LCC HD9506.P262 S57 2020 (print) | LCC HD9506.P262 (ebook) | DDC 338.2/741099575—dc23

LC record available at https://lccn.loc.gov/2020016102

LC ebook record available at https://lccn.loc.gov/2020016103

British Library Cataloguing in Publication Data

A catalogue record for this book is available from the British Library

ISBN 978-1-78920-856-6 hardback
ISBN 978-1-78920-857-3 ebook

In memory of Józef Skrzypek –
an extraordinary man

Contents

Illustrations, Figures, Maps and Tables

Illustrations

Figures

Maps

Table

Preface

Pastor Jacob asked me to join him as he sat, cross-legged, on the floor in his daughter's house in Paupe – a so-called impact community of the Frieda River Project. He said that he would like to tell me a story, and he asked that I record it carefully:

> My father, Aiyapanai, was a young husband when a white woman and her team travelled up along the Sepik River. Once they arrived in Iniok, the white woman looked up the river flowing into the Sepik from the south and asked: 'Who does this river belong to?' One of the local men, Komnei, stood up and said: 'This river belongs to me'. The Iniok group, they call this water Inayap. We, the Paiyamo, we call it Neigeme.
>
> The white woman stayed in Iniok for some time. She used that time to recruit local men from Iniok and the nearby village of Tauri and asked them to assist her on her onward journey, up against the currents of Neigeme. They set off on pull dugout canoes and after a week or so they arrived at Ulibi – ground belonging to Samuel Nano – not far from the current location of Aston. Once there, they dug the ground and collected stone samples. Then they came this way, to Amumuiu. They continued their work in a place we call Piamoku, then went down towards Arinou, the last Melefale mountain, before entering Ulibi. Aiyapanai, Nano and Joel's father Utawane were in the group of local men who came to the river banks to greet the new arrivals. It was a Nasuna group. They did not speak English, and they did not speak Tok Pisin;[1] they only spoke the local language – Paiyamo Amoke. At that time, Ayapanai was already married and had two children, Sibaku and Sruamo, father of Peter. But that day, he only took Sibaku with him. The white woman gave them rice, matches, salt and soap. It was the first time they had tried rice – they didn't like it and they threw it away. It was also the first time for all of them to see white skin and they were scared. Back then they all lived at the back of Ulibi, two, three hours walk away from Amumuiu. Two other Paiyamo groups, Inaku and Nesiar, lived further away. They did not meet the white woman.
>
> The woman and her team dug the ground on the banks of the river, although Aiyapanai didn't know why. When they finished their work, they left, never to return. But the woman did complete a report from the visit. She was German, and it was her who renamed Neigeme, giving the river her own name – Frieda. It was 1942.

Twenty years later, in 1962, a kiap[2] arrived in this area following the Frieda River from the Sepik. He brought everyone together into one village and gave the village the name Paupe. Back then, Paupe was located on the land which belonged to the Inaku group – a fair distance down the river from today's village site on the Melefale land.

In 1969, a company called CEC came to Frieda. They found the German woman's report and map and came. They arrived in a helicopter and landed in Paupe. A kiap and a policeman were the first people to leave the chopper and approach Aiyapanai. The kiap asked: whose ground is this? When the kiap asked, Aiyapanai nodded his head. He didn't understand the language of the new arrivals. The kiap then wrote down Aiyapanai's name – he thought that the man was saying yes and allowing the company to come and work here. The people who arrived on the helicopter went to Melegeme where they collected stone samples for testing. And that stone was very good. That stone belonged to Melegeme. Now everyone calls that place Nena.

And so in 1969 the exploration work started. The drilling commenced in 1970 and continued throughout 1971. They came here in 1969. And they are still here. And so are we.

Pastor Jacob finished his story by saying:

I became a leader because my father was a leader before me, and he was the one who, on behalf of Melefale, told the company: 'you may come in'.

My father opened the doors for the company to come here. He only spoke his language, the Paiyamo Amoke, so he only nodded when they asked him. Now I am the leader and I speak Tok Pisin. My son, Aiyap – he will be the leader after me. And he will know how to speak English.

Notes

1. Tok Pisin is the lingua franca in Papua New Guinea.
2. Colonial patrol officer.

Acknowledgements

This book would not exist without the Paiyamo. My deepest gratitude goes to pastor Jacob Aiyapanai, Lina, Dinah, Mawi, Rhonda and my Melefale family to whom I owe a debt of gratitude for the care they extended to me, their support for my work and their friendship. People in Paupe community welcomed me into their lives and patiently and generously encouraged me as I tried, at times very clumsily, to understand their world and their role as an impact community of the Frieda River Project. I learned a great deal from them and feel honoured and humbled to have been able to get to know them and their places. I am unable to name all of them here – but I thank each and every one of them.

I would not have been able to complete this project in this scope if it wasn't for the participation of the company's Community Affairs personnel and people representing provincial and national governments at the Frieda River Project. I thank members of the Community Affairs team, and the team manager, for their trust, support and for their generosity in sharing their experiences with me. I also want to thank Papua New Guinea's National Research Institute for their assistance and support during fieldwork.

This book is based on research conducted at the University of St Andrews' Centre for Pacific Studies under guidance of Dr Tony Crook. The manuscript was prepared during a two-year secondment to the University of Queensland's Centre for Social Responsibility in Mining. The work presented here was made possible by financial support in the form of an ESRC studentship, The Russel Trust travel grant, and funding from the European Union's Horizon 2020 research and innovation programme under the Marie Sklodowska-Curie grant agreement No. 753272. I would like to thank Dr Crook and my colleagues at both centres for helping me test and develop my ideas about a mine at Frieda. I am particularly grateful to Assoc. Prof. Nick Bainton and Prof. John Owen for their encouragement and their insightful and invaluable comments on the draft versions of this book. I would also like to extend my thanks to Dan Jorgensen for all his help and ethnographic guidance; Thomas Moylan and Tony Friend for generously sharing with me their field reports, which were a great source of knowledge and information; and Roger Lohmann for his detailed feedback on a draft of chapter 3.

None of this would be possible without my family and friends, whose love and support I felt every step of the way. I want to thank all of them for sharing this experience with me, for trusting and supporting my ideas and choices and for being a bottomless source of strength and inspiration. Most of all, I want to thank my parents, Elżbieta and Marek, my sister Janka, and my partner David. I am forever grateful for their unconditional love and for being truly happy for me throughout this project, even though I know how many sleepless nights it has caused them. I would not be where I am and who I am without them.

Notes on Text

A number of Tok Pisin terms have been used in the manuscript. Tok Pisin is the lingua franca in Papua New Guinea. Terms that have only been used sporadically are explained in the text. Terms that are used repetitively are explained here.

- *Bilum* – a traditional string bag.
- *Kastom* – the term derives from the English word 'custom' and is a highly politicised concept in Papua New Guinea. It tends to be used in reference to traditional beliefs and practices; however, its concerns are very much contemporary.
- *Kastom lain* – a collective term used to describe men who have been initiated into the traditional cult and were involved in the management of traditional religious knowledge.
- *Kiap* – colonial patrol officer during Australian administration
- *Haus tambaran* – men's cult house, traditional site of worship.
- *Masalai* – a supernatural being, commonly associated with particular places and often explained as a land spirit.
- *Papagraun* – term traditionally used to describe individuals with a claim over particular areas of land, increasingly used in a communal sense to mean 'landowners'.
- *Sanguma* – sorcery.

In the text, the Frieda River Project is referred to either by its full name or as 'the Project'. Xstrata Frieda River Ltd is generally referred to as 'the company' except where there is a need to differentiate it from other companies. In those cases, it is simply referred to as Xstrata. Following terminology used by my interlocutors, I use the term 'Frieda' to describe the Frieda River area, location of the Frieda River Project.

Introduction

This book aims to enhance our understanding of the complexities of the Frieda River Project in Papua New Guinea (PNG). It tells a story of local stakeholder strategies on the eve of industrial development, told largely from the perspective of the Paiyamo community of Paupe – one of the Project's so-called 'impact communities'. It is a story of what Pijpers and Eriksen (2018: 2) glossed as 'mining encounters', understood as 'the negotiations and frictions between individuals and groups with different agendas, world views and aims within the context of mining operations'. Engaging local ideas of knowledge, belief and personhood, it explains how fifty years of encounters with exploration companies shaped the Paiyamo's aspirations and made them revisit and re-examine the past and develop new strategies to elicit a way of moving towards a better, more prosperous future.

Although the Frieda River Copper-Gold Project has not yet even obtained a mining permit, it has already reached national acclaim as a contentious resource development project. The Project is located on the boundary between the country's East Sepik and Sandaun Province, some 200 km inland south from the Pacific Ocean coast and 75 km east from the Indonesian border. By any contemporary standard, its location is very remote – amid tropical rainforest and in the fragile ecosystem of the upper reaches of the Sepik River. Major civic infrastructure in the Project area is almost non-existent, and the provision of government services is near absent.

Rich copper and gold deposits were first identified there in 1966/67 during a regional mapping exercise. In 1969, a Prospecting Authority (today known as Exploration Licence 58, or EL-58) was granted to an Australian mining company, Mount Isa Mines Ltd, and its subsidiary, the Carpentaria Exploration Company. Exploration activities began shortly after and have been ongoing since. They markedly intensified in the mid-1990s when, in the middle of a resource boom, Highlands Gold Pty Ltd completed the first full feasibility study for the Project. That study cast great doubt on the economic viability of the venture, placing it among a special cohort of projects worldwide known colloquially as 'stranded resources'. In 2007, Xstrata

Frieda River Ltd (also referred to here as Xstrata or 'the company') bought into the venture with the aim of completing another feasibility study for what was then referred to as the Frieda River Project. Xstrata Frieda River Ltd was a PNG registered company with headquarters in the country's capital, Port Moresby. The company was embedded within a much larger, Australia-based parent institution, Xstrata Copper, which at the time was the fourth largest copper producer in the world. The level of on-site activity at Frieda River intensified and remained high until December 2011 when the company completed the Environmental Impact Statement (EIS) for the Project.[1]

The chapters in this book are based on ethnographic fieldwork conducted in 2011 and 2012 – spanning the build-up of expectations ahead of the anticipated submission of the EIS by Xstrata to the PNG government and the period of uncertainty that followed the company's subsequent decision not to file the document. Members of local 'impact communities' saw this decision as a sign of the company's intention not to proceed with the Frieda River Project. Their intuition was right. In 2013, Xstrata Copper merged with another mining company, Glencore, forming Glencore Xstrata – the world's fourth biggest mining company. In 2014, a significantly smaller Brisbane-based company, PanAust, entered the project as the main developer.[2]

The Frieda River area (also referred to here as Frieda) is home to one of the largest undeveloped copper and gold ore deposits in the Pacific. It is estimated that its three deposits, Horse-Ivaal-Trukai, Ekwai and Koki, contain more than 2.6 billion tonnes of copper and 0.24 grams per tonne of gold. Current predictions approximate that, once operating, the mine will produce an average of 175,000 tonnes of copper and 235,000 oz of gold a year (peaking at 290,000 tonnes and 270,000 oz respectively), with an estimated lifespan of 33 years (Coffey 2018a: 2). Recent analysis by Valenta et al. (2019) included Frieda's mineral assets on a list of the world's most complex copper orebodies, where this complexity extends beyond geological characteristics of the deposit and includes a range of environmental, social and governance factors. The Frieda River Project exhibits a number of characteristics used by Valenta et al. to define and identify what they call 'complex orebodies'. The size of its copper reserves makes Frieda River a world-class deposit, but the grade of the mineral is relatively low, meaning that any future extraction will require significant input of water and energy and will likely produce high volumes of waste, resulting in a large environmental footprint project. The deposit is located in a remote and ecologically sensitive area; on land belonging to indigenous communities with limited access to basic services and very poor civic infrastructure and in a country grappling with poverty and corruption. Valenta et al. argued that although risks associated with environmental, social and governance factors may not be price-sensitive in the immediate term, rapid unlocking of these ore bodies to meet global demand for minerals will unleash some very complex impacts and can threaten the future of individual projects as well as global supply of copper.

What Is a Mine?

Understanding what 'a mine' is in these contexts and what it means to different people is critical to recognising and comprehending complexities of the Frieda River Project. In her seminal work, *The Gender of the Gift*, Marilyn Strathern (1988: 13) challenged the antinomy between 'society' and the 'individual' as key social categories in Melanesia and argued that Melanesian persons are not 'conceptually distinct from the relations that bring them together'. Instead, they 'contain a generalised society within. Indeed, persons are frequently constructed as the plural and composite site of the relationships that produced them' (ibid.). In the field of anthropology, many scholars engaged with this dividuality, taken as an alternative model of personhood placed in opposition to the 'Western individual' (see, for example, Smith 2012 and Mosko 2010a, 2010b). My reading of Strathern is that her main objective was not to provide us with a self-contained and distinctive model of a Melanesian person but to challenge us to imagine what a social theory would look like if it took something other than the 'society' and the 'individual' as basic units of analysis. In *The Gender of the Gift*, Strathern (ibid.: 309) herself explains:

> I have not authored 'a perspective' on Melanesian society and culture; I have hoped to show the difference that perspective makes, as one might contrive an internal dissociation between the character of an author and the character of his/her characters. Consequently, I have not presented Melanesian ideas but an analysis from the point of view of Western anthropological and feminist preoccupations of what Melanesian ideas might look like if they were to appear in the form of those preoccupations.

Through her ethnographic and intellectual inquiry into dividuality, Strathern skilfully demonstrated how the change in perspective forces us to reconsider how we think about people, the world and the different elements within it. Pivoting on these ideas, in this book I want to challenge how we think about 'a mine' in the context of resource exploration ventures, where in the absence of mining infrastructure and industrial extraction activities planning processes are characterised by prolepsis grounded in divergent world views, knowledge systems and experiences of individuals and groups on different sides of resource relations. Schram (2015: 319) argued that the concept of a dividual person 'was never meant to be a theory of particular cultural norms. It was always a thought experiment about what social theory would be if it started from an alternative understanding of persons'.

This book presents another thought experiment, one which explores what a mine would look like if we took something other than the physical form of a mining pit and industrial infrastructure as a starting point of analysis. My aim is to create within these pages a discursive space through which, despite the lack of physical appearance commonly associated with industrial resource extraction projects, the mine at Frieda can be brought to life and made visible. To that end, the book traces the ways in which the Paiyamo came to understand and know 'the mine', which, in the absence of visible physical mining infrastructure, they came to experience in terms of social relations, community and individual aspirations, and novel mechanisms for affecting an array of future opportunities.

Those opportunities formed a particular kind of imagery in Paupe, one through which community members came to associate the mine with broadly conceived 'development' and new kinds of livelihoods this 'development' would bring. This imagery was shaped by a long and varied history of encounters with corporate and state agents drawn to Frieda by the prospect of mining and is informed by a range of other factors such as communities' knowledge of other mining projects, their visits to urban centres, gossip surrounding their reading of national newspapers and images from predominantly American movies they watched on pirated DVDs from Indonesia during rare movie nights in the village. In very general terms, they hoped that, alongside financial payouts, once operating the mine at Frieda River would bring a range of services, education, training and employment opportunities and physical as well as communication infrastructure. Reflecting the importance of relationality in Melanesian societies, the book is loosely framed around Marilyn Strathern and Roy Wagner's ideas about relationality as a tool for transformation (see chapter 3). It follows the Paiyamo in describing the transformative potential they saw in forming relations with representatives of extractive companies arriving in the area and the strategies they developed to shape those relations to affect change and bring about much anticipated development.

Life Cycles of the Frieda River Project

There are multiple histories of resource development at Frieda. These reflect the divergent ways in which different individuals and groups 'do knowledge', assess evidence and understand and get to know the mine. The extractive companies tend to describe the history of the Project in terms of feasibility and scoping studies, based on the record of changes in corporate ownership of the EL-58 and activities undertaken by consecutive companies holding a stake in the Project (see chapter 1), and in relation to the different stages of various administrative exercises characterising the process of planning, constructing and operating a large scale resource extraction project. In Papua New Guinea, those projects progress through what is referred to as a 'mineral project life cycle' (Figure 0.1.) set out by the Mineral Resources Authority – a government agency responsible for regulating mining activities in the country. Each phase of the cycle is aligned with particular licensing requirements and approvals that legally frame key relationships and obligations between the company, the state and, to an extent, local communities. Shortly after I first arrived at the Frieda River Project site in 2011, a member of the company's Community Affairs team sat me down and explained to me that despite over forty years of exploration Frieda River was not yet a mine but remained a resource extraction project at the exploration and feasibility stages of its life cycle. 'This is a project, not a mine, and we need to make sure that we are very clear about that,' he told me. In this perspective, it is only once the mineral resource project passes into the latter stages of its life cycle when the completion of a series of legally binding approvals opens up space for construction of infrastructure and extraction of the mineral resource that a project becomes a mine. To a large extent, in corporate and legal terms, it is this progression from a project to a mine that deems the resource extraction venture viable and the relations it entails binding.

At the time of my fieldwork, the fact that the decision regarding the future of the Frieda River Project (also referred to here as the Project) had not yet been made had a strong impact on the company's activities and ways in which it engaged with the Papua New Guinean and provincial governments, and local communities. Although often described by the company in terms of 'maturing stakeholder relations', these engagements were framed in a way that reflected the logistical and, crucially, financial uncertainties regarding the future of the Frieda River Project, viewing the mine as a possible but tentative end result of a complex geological, technological, legal, political and financial exercise. At the time of research, the company's payouts to the local communities were limited to periodic compensation payments for the disturbance of land used for either drilling or company infrastructure and

Mineral Project Life Cycle
Licensing Phases

Exploration License		Development License (Mining Lease, Special Mining Lease)			Closure - Relinquishment		
Exploration	Feasibility	Approvals	Construction	Operation	Decomposition and Rehabilitation	Post Monitoring	Relinquishment

Figure 0.1. Mineral Project Life Cycle Licensing Phases adapted from Mineral Resources Authority Licensing Information available online at http://mra.gov.pg/Portals/0/Mining%20Info/licensing%20phase.pdf. Public domain.

irregular payments generated by often ad-hoc and short-term employment opportunities. However, not discouraged by these limited financial rewards, and driven by the opportunities they envisaged in future mineral development at Frieda, the Paiyamo actively engaged with the company representatives and regularly demonstrated their support for the Project. The way in which they spoke about 'the mine' differed significantly from the corporate view of the mine as a physical demonstration of a sort of maturity of geological and economic factors that placed the Project at a particular point of the mineral project life cycle. Rather, they saw the mine as a demonstration of relationships they developed and maintained over time with the company developing the project, other impact communities, the national and provincial governments and a number of supernatural agents that resided in the local landscape – volitions that may not be visible, but that can alter the physical reality and the world (see chapter 3).

For the Paiyamo, uncovering the history of Frieda's mineral wealth and the uncovering of Frieda's rich copper and gold deposits were, in many ways, mutually constitutive processes, framed around particular kinds of knowledge and relations made visible. In the local universe of constantly negotiated terms and relations, and in the context of the absence of mining infrastructure and industrialised mineral extraction activities, establishing the meaning of the term 'mine' became, as the following chapters will show, an ontological undertaking. While in official records the history of the Frieda River Project begins in 1966/67 when the area's copper mineralisation was first formally identified, the Paiyamo traced its origins to a much earlier event. In an attempt to explain to me the history of the Project, the Paiyamo told me the following story:

> The company is working to find gold and copper at Frieda. But what they are really looking for is a stone we and the Telefols call *bangri*.[3] This story belongs to the Telefols, but it is our story, too.[4] Before she left Telefolip, the old woman Afek[5] sent four spirits with four stones to the different parts of her area. She asked the spirits to hide the stones – and to hide them well. One of the spirits went to the Western Province; another one to the Highlands. The third stone was a woman dressed in traditional clothes, with grease on her skin. She too was sent to the Highlands. The fourth spirit came to Frieda and decided to hide his stone here. Now we know that the four stones were mineral deposits. The one which was hidden in the Western Province is now the Ok Tedi mine. The one in the Highlands is the Porgera mine. The third stone, the lady, is the oil deposit which is being explored in the Highlands. The fourth one is Frieda. The stone is hidden by spirits called the Norenmin. Sometimes, during night shifts, people working on the exploration sites can see them. If you are hunting in the bush around that area and hear a cracking sound, it's them. They carry special *bilums*,[6] decorated with feathers, pig tales and traditional money. If you can kill one, and take their

> *bilum*, the *bilum* will provide you with many special powers. But in the past, all people who have heard the sound of Norenmin came back home and died. It is an ancestral story, but now things are happening and the story is becoming true. It is an old story but it is true now. The spirit was hiding the stone from many companies. This is why they did not extract the copper and gold here. But it does not hide it from Xstrata. Before we did not want the mine, but now we know how important development is, so we asked the spirits to stop hiding the stone.[7]

The myth of Afek and *bangri* is just one example from an elaborate repertoire of stories that the Paiyamo used to explain the Frieda River Project to themselves, the government representatives and the company's on-site personnel. In these myths, stories of the ancestors intertwined with stories of corporate and state agents, biblical narratives and local genealogies. They were deeply embedded in the local aesthetic of revelation in which events are never fully knowable at the time of happening – their significance and effect are revealed over time.

Anthropological literature about mining in Papua New Guinea is vast. For the most part, it discusses resource extraction projects through a prism of conflict and conflicting ontologies (see, for example, Filer 1990; Connell 1991; Kirsch 1997; Wardlow 2004; Reilly 2008) as mineral projects by their very nature tend to bring together various people and groups with some inherently different ways of experiencing the world. Most often, the accounts focus on cultural misunderstandings and their consequences as they unfold in the advanced stages of the mineral project life cycle, when and after a particular mine has reached its 'operational maturity' (e.g. Golub 2014 and Jacka 2015 on Porgera; Kirsch 2014a on Ok Tedi). But there has been very little ethnographic research that has focused explicitly on the anticipation of possible but uncertain future resource development. Exploration projects at greenfield sites, such as the Frieda River Project, stand out against already operating mines. They embody the raw uncertainties of resource extraction, where exploration activities raise expectations of mineral development whilst only a fraction of known deposits become operating mines. They tend to be capital intensive while bringing in no income, and they require development of an extensive body of knowledge about their geological but also environmental, social and governance setting (see chapter 1) before they can proceed. The footprint of exploration activities on the natural environment tends to be limited to the impact of scoping and testing exercises required for assessment of a project's feasibility and, in more advanced stages, the most appropriate project design. The impact on the social environment, however, tends to be significant.

The positioning of the Frieda River Project at an early stage of the mineral project life cycle creates an important and unique research space, located somewhere between experiences of a present reality and anticipation of an imagined future. Encounters between exploration companies and local communities introduce new visions of what this future might look like and, through complex systems of relations and administrative processes, create new pathways of how to actuate that future. One of the most detailed studies of this process in the Papua New Guinean context comes from Monica Minnegal and Peter Dwyer.[8] In their recent book *Navigating the Future: An Ethnography of Change in Papua New Guinea*, they (2017: 270) describe how the Kubo people in the country's Western Province navigate what the authors refer to as 'the ever-shifting landscape that lies between present circumstance and present desire' to affect and effect an uncertain future, informed by the proximity of the large liquefied natural gas project (the PNG LNG project operated by ExxonMobil PNG). Minnegal and Dwyer describe how 'outsiders', including exploration company personnel, enhanced the desire and expectations of the Kubo, presenting them with new models of what could be possible. Their book paints a picture of the Kubo as agents actively involved in shaping their imagined future but who have not yet mastered the mechanism of how to get there. A version of this process unfolds in Paupe, where decades of encounters with extractive companies have shaped the community's ideas about the future and how it can be brought about. These are embedded in local ideas about relations, personhoods, knowledge and change – all of which will be discussed in the chapters that follow.

Local Context

The Frieda River Project is located in a mountainous area on the Northern slopes of the Thurnwald Range in north-western Papua New Guinea. The altitude varies between 50 and 1,700 meters. The Project area is covered almost entirely by tropical rainforest, with few signs of disturbance to the vegetation either by local communities or the successive exploration companies, and it is characterised by higher than average diversity of terrestrial fauna (Coffey 2009: 15) – living up to its status as a greenfield site development. In general terms, the climate there can be described as wet and tropical, with frequent storms and high year-round humidity. The area is very remote, and transport infrastructure connecting local communities to larger urban centres such as Wewak or Vanimo is scarce. Recent plans for the venture describe the Frieda River Project as 'a core component of a transformative nation-building opportunity – the Sepik Development Project', that promises to deliver broad socio-economic uplift, 'improve transport and connectivity, facilitate the delivery of services to remote communities, and boost domestic and international trade and investment' (PanAust 2018: 1). The

Project's current operator has included in their plans, alongside the Frieda River Copper-Gold Project, the Frieda River Hydroelectric Project, which encompasses an artificial reservoir and a 600 MW hydroelectric facility; the Sepik Infrastructure Project involving construction of a road connection between the Project site and Sandaun's provincial capital Vanimo as well as an upgraded Vanimo Ocean Port and Green River Airport; and the Sepik Power Grid Project, comprising a 370 km long transmission line providing power to Project facilities in Green River and Vanimo (with the possibility of excess power being distributed to customers in Papua New Guinea and Indonesia) (Green and Crouch 2018: 1). For the time being, however, the Project's footprint has been condensed to impact a relatively small population with recognised land ownership claims in the Project area (see Map 0.1 below). In 2009, the total population of those communities was estimated at just under 1,500 people living in 218 households. By the 2017 Project census, this population had increased to just over 2,000 people, and the number of households reached 310 (Coffey 2018b: 11).

The history of the area is long and complex, characterised by genocidal war and conquest, which began in the late 1800s. This period of local warfare lasted until the 1950s and 1960s when the first colonial patrols and missionaries reached this corner of New Guinea. As described by lawyer-anthropologist Jim Fingleton in a report he produced for Xstrata:

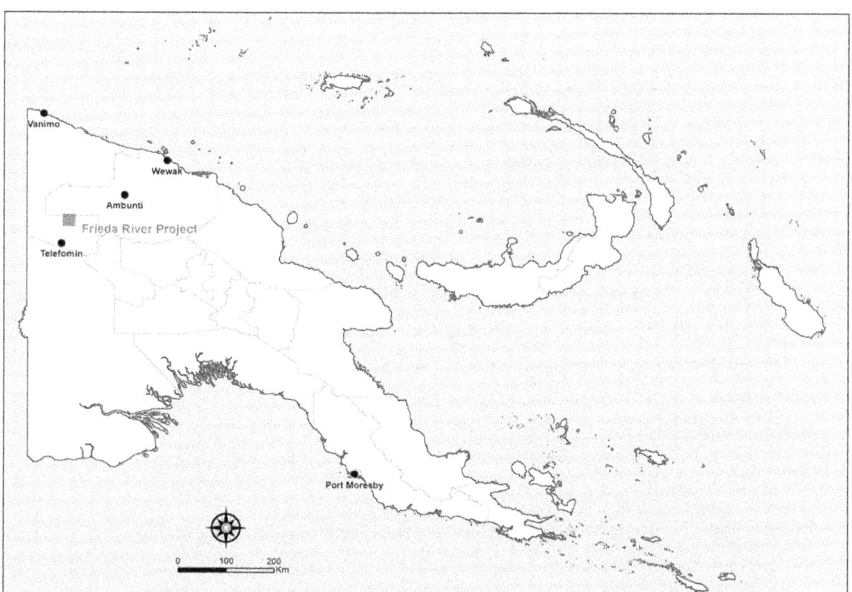

Map 0.1. Location of the Frieda River Project. Map created by Kamila Svobodova.

Map 0.2. Frieda River Project Area. Map created by Kamila Svobodova.

> Both geographically and culturally, the project area lies in a 'transition zone', between the larger populations inhabiting the high peaks of the New Guinea cordillera to the south and the extensive plains of the Sepik basin to the north . . . Over the centuries, the project area has seen dramatic shifts of population, as Mountain Ok groups from the south expanded into the area, wiping out previous occupants. The invaders raided and killed, capturing women and children, who were adopted and incorporated into their communities as they progressively settled the area. Groups were hungry not for land but for people. Their incorporation of captives has led over time to the extraordinary complexity of the groups presently living in the project area. (2010b: iv, quoted in Green and Crouch 2018: 27)

In his report, Fingleton estimated that in the past the Project area was occupied by representatives of some twenty different cultural groups. By the mid-1950s, however, most of those had been either extinguished or incorporated into others. This history of warfare has an impact on landowner identification processes for the Frieda River Project (see chapter 4) and continues to influence local and regional politics, with stories of conquest used in landowner identification exercises alongside ancestral connection to land as evidence of claims over land and its resources.

Currently, inhabitants of the planned mine area include representatives of three distinctive language groups: the highland groups of Miyanmin and Telefomin, and the Paiyamo who live in the transition zone on the foothills and plains along the Frieda River. The Miyanmin and the Telefomin belong to the Mountain Ok family of languages and share a common mythical ancestress, Afek (see Craig and Hyndman 1990). Historically, they both have made successful claims to the ownership of land covered by the Special Mining Lease for the Frieda River Project (see chapter 4). This means that if the Project proceeds to operations and becomes profitable they stand to receive royalties as well as compensation payments and other forms of financial benefits. The Telefomin are considered one of the most numerous and politically powerful language groups in the region, famed for their warfare skills and extensive conquests. The Paiyamo, on the other hand, is the least numerous group, having been hugely reduced during the period of warfare (see chapter 2). Linguistically, the Paiyamo are related to the Wario and Leonard Schultze groups to the east, and the Duranmin to the south, rather than the Mountain Ok communities to the north-west. Through intermarriage, they are further connected to Eliptamin, Duranmin, Telefomin, Wario as well as groups living in the May River area, and historically trade ties placed them within even wider regional networks. In 2011/12, they were considered landowners not in the Special Mining Lease area, but in the area designated under the Lease for Mining Purposes. This meant that while they were not considered owners of the land immediately above the deposit, their land was of crucial importance for the Project infrastructure. Their position as an impact community was well grounded, and their level of involvement with the Project was consistently high.

At the time of research, seven villages held the impact community status and were actively involved in the processes taking place under the umbrella of local stakeholder engagement (see chapter 1): Miyanmin (henceforth Miyan or Miyans) communities of Wameimin 1 and 2, Sokamin and Amaroumin; Telefomin (henceforth Telefol or Telefols) communities of Ok Isai and Wabia; and a Paiyamo community of Paupe. In the field of anthropology, the Miyans have been studied extensively by George Morren, whose work focused on human ecology and environmental change (1979, 1981, 1986), and Don Gardner, who produced detailed accounts of the Miyan settlement and residential patterns (1980), social organisation (1981) and ritual technology (1983), as well as belief systems and conceptions of agency (1987). Alongside two other anthropologists working in the area, Dan Jorgensen (with the Telefols) and Rune Paulsen (with the Iwam people of May River), Morren and Gardner were involved in a social study of the Project impact area commissioned in the mid-1990s by Highlands Gold – a company managing the Project at the time. Results of the study were used in a landowner identification exercise, discussed in more detail in chapter 4. As recognised landowners in the Project area, the Miyan communities play an important role in the process of planning for a mine at Frieda, and this role needs to be acknowledged here. However, the Miyan communities are not discussed in detail in this book. Instead, the research presented here is focused predominantly on Paupe and the Paiyamo, the smallest and least researched group of the Frieda River Project,[9] with some supplementary material gathered in Ok Isai and Wabia – the Project's two Telefol communities.

Paupe, Ok Isai and Wabia are all located on the banks of the Frieda River, and my informants sometimes referred to them collectively as 'the Frieda community'. Family links built through intermarriage meant that the interactions between the three were common and frequent, and my research in Paupe inevitably led to numerous visits to Ok Isai and a few visits to Wabia. Since the start of mineral exploration at Frieda, all three settlements have moved from their previous sites, some more than once, relocating closer to the river in search of better access to transport routes and services. Anticipation of benefits from the Frieda River Project played a major part in those moves, as community members sought access to ad-hoc employment opportunities, services either provided or facilitated by the project developer, and recognition as stakeholders of the Project and the future mine. In their current locations, the three villages are within easy access of one of the company's main residential camps, the Airstrip Camp, located a short walk south from Paupe, and a motorised canoe ride north from Wabia and Ok Isai.

Travelling between Paupe and the two Telefol communities, canoes cross a provincial border. Six out of the seven impact communities of the Frieda River Project (Ok Isai, Wabia, Wameimin 1 and 2, Sokamin and Amaroumin) belong to Sandaun Province (Telefomin Rural LLG) – the north-westernmost province of Papua New Guinea with a provincial capital in Vanimo. The location of the mineral deposits falls within Sandaun's administrative borders, which has led many (including representatives of provincial administration interviewed for this study) to view the Frieda River Project as an exclusively Sandaun project. Paupe, on the other hand, belongs to the East Sepik Province (Tunap-Hunstein LLG) located to the east of Sandaun, with a provincial capital in Wewak. Numerous changes to the design of the Project have meant that the East Sepik Province's importance in terms of hosting Project infrastructure has varied significantly over time. However, the Project's location in the tributary of the Sepik River means that, whatever the final design, while the benefits will accrue in Sandaun, the risk is shouldered by the East Sepik Province. In case of an industrial incident leading to contamination of the river system, it is the East Sepik and not Sandaun Province that will be most affected.

The levels of opposition to the Project are significantly higher in communities living in the Sepik River basin downstream from the Frieda River than in the Project's immediate locality, with some politicians as well as community groups and members publicly calling for the Frieda River Project to be halted, often comparing the project's natural setting to that of Fly River in Western Province, made infamous by the environmental disaster caused by a series of industrial accidents and environmental law breaches at the Ok Tedi mine (see Kirsch 2014a). The 2019 report prepared by the Australian-based Jubilee Australia Research Centre in collaboration with the Papua New Guinea-based Project Sepik painted a picture of the Sepik communities as being largely left out of the conversations with the Project developer. It outlined some of the risks associated with the Frieda River Project and concluded that 'the potential for this project to lead to damaging social conflict and unrest is real and must be taken seriously' (Fletcher and Peni 2019: 5). In other narratives and considering the very limited infrastructure and government service delivery, the Project is described as a much-needed pathway to bring development and services to the province. As an East Sepik provincial government representative explained to me: 'we are very remote here, we don't have development. [The national] government does not help us. With the mine, the company will bring the money for development. We understand the problems associated with mining projects. . . . But if we don't get major development in the province, where do we get the money to help people?' This statement reflects a common sentiment in Papua New Guinea – one that continues to see the resource extraction industry as a pathway to fulfilling national and regional development aspirations. This is despite the impact of the country's infamous mining projects such as the aforementioned Ok Tedi and Panguna mines.[10]

National Context

The role of the mining industry in Papua New Guinea's political and economic strategies post-independence provides important context for the Frieda River Project and the primary research material presented in this book. Following almost seventy years of Australian colonial administration, the country gained independence in 1975. Papua New Guinea emerged with no particular sense of national identity apart from an unevenly shared colonial history (Jorgensen 2007a) and external symbols of nationalism that bore little actual meaning for most of its people (Power 1997: 84; see also Hirsch 1990). However, as noted by Jorgensen, 'what Papua New Guinea lacked by way of common tradition at independence was more than made up by an enthusiasm for development in all regions of the country, and many of the new state's claims to legitimacy were based on the promises that all Papua New Guineans could expect development to come their way' (2007a: 58; see also Hirsch 1999). As it withdrew, the colonial administration left a void in the provision of services and in the maintenance and development of infrastructure that the newly formed government of Papua New Guinea found difficult to fill. In the years and decades following independence, the country's public services were functional yet limited in their capacity to respond to the needs of the country's growing population (Gelu 2010), meaning that in large parts of the country the civic infrastructure fell into disrepair and access to government services diminished.

At independence, the new government's ideas about development were based around a largely egalitarian rhetoric as outlined in the country's Eight Point Plan (1975). By the 1980s, tensions arose between the goals of 'development for all' and the realisation that development, by its very nature, produces inequality (see, e.g., May 1998; Rivers 1999; Hirsch 2001b; Stead 2017). These tensions intensified further when the country turned its hopes for development towards large-scale mining projects, creating new boundaries of influence and emphasising 'the new priority of economic over social goals and political relations' (Hirsch 2006: 154). The government embraced the promise of resource-driven development and, utilising the resource boom of the late 1970s, actively worked to strengthen the country's position as a 'mining resource frontier' (Hyndman 2001). Over the decades that followed, mining became a central feature of Papua New Guinea's growing economy, capitalising on what Ballard and Banks (2003) referred to as the 'mining exploration bonanza' of the 1980s and early 1990s, which saw intensification of exploration activities in many parts of the country, including the copper and gold prospects at Frieda. By the end of the twentieth century, 156 Mining Exploration Licences had either already been awarded or were awaiting grants, and four major mines (Ok Tedi, Lihir, Misima and Porgera) and one medium-sized mine (Tolokuma) were in operation (Hirsch 2001c).

In line with the prevailing development paradigm of the time, the mining industry seemed to offer an ideal opportunity to bring revenue to the national economy and the desperately needed development into the local areas (Gilberthorpe and Banks 2012). Over time, 'in the minds of most Papua New Guinean "grassroot" or village people, mining has become the way to gain wealth rapidly and to ensure that dreams of "development" and "modernity" come true' (Filer and Macintyre 2006: 216). This subsequently created 'unparalleled expectations of mine developers to provide services where there is a lack of government presence' (Imbun 2007: 188). Literature on the subject has made links between mining prospects and what has been referred to as a 'cargo cult mentality' (see Lattas 1998; Stewart and Strathern 2000b; Jorgensen 2005; see also chapter 3 here) and argued that the vast majority of people in PNG 'now seem to subscribe to the belief that their land does contain some valuable resources – whether gold, oil, diamonds, or the truly visible logs – and that their only chance of development lies in their share of the rent to be collected from the extraction of these resources by some multinational company' (Filer 1997b: 113; see also Leach 2004). Evidence shows, however, that 'indigenous peoples have historically gained little from large scale resource development on their traditional lands and have suffered from its negative impacts on their cultures, economies and societies' (O'Faircheallaigh 2008: 1). Mining has become 'central to the evolution of the notion that resources can be a curse that gives rise to a lack of development, internal tensions, human rights abuses, and conflict at the national level' (Ballard and Banks 2003: 295). Despite these negative impacts, which are well publicised in PNG, the local enthusiasm for the Frieda River Project is high.

The resource curse has been linked to a governmental (bureaucratic and administrative) incapacity to distribute benefits to its people (Jackson 1991); a disconnect between decision makers and the society (Collier 2008); a preoccupation with distant, international opportunities rather than locally defined social, economic and cultural goals (Connell and Howitt 1991); and the growing evidence that mining projects tend to limit, not expand, economic alternatives both at the local and national scale (Gerritsen and Macintyre 1991). In Papua New Guinea, the resource curse has been associated with a phenomena described by Filer et al. (2008: 175) as a responsibility vacuum, which 'makes partnerships between the corporation, local community and the government difficult, dangerous, or downright impossible' as none of the agents involved are prepared to take full responsibility for their own actions. Although the empirical evidence for the existence of the resource curse is difficult to pin down, what evidence is available clearly points to the dangers of resource-based, monopoly economies (Sachs and Warner 2001) in which 'dependence on natural resource rents impedes the development and proper functioning of state institutions' (Genasci and Pray 2008: 38). This is amid increased realisation that the negative force of the resource curse is not a purely economic problematic, but it can 'also be portrayed in political, cultural, or environmental terms, which means that the resource curse may have several dimensions in the same country or region' (Filer and Macintyre 2006: 217). The multifaceted nature of the role of extractive industries in national and local development is reflected in Valenta et al. (2019) modelling of complex orebodies. In this model, environmental, social and governance factors need to be considered alongside geological and technical characteristics in assessment of project feasibility and impacts and, ultimately, in making the decision on whether or not the given project should go ahead.

The Beginning

The first time I came across the Frieda River Project was in a letter exchange published in *The National*, one of Papua New Guinea's two main daily papers, between Andrew Moutu, an anthropologist and activist from Port Moresby, currently the director of Papua New Guinea's National Museum; Greg Anderson, then Executive Director of Papua New Guinea's Chamber of Mines and Petroleum; and Dugi Wilson, the then General Manager of the Xstrata Frieda River Project. Dr Moutu's letter warned against the possibility of the Frieda River Mine turning into an ecological disaster of great proportions, threatening ecosystems as well as livelihoods of people living in communities on the Sepik River, downstream from the proposed site of mining activities. Mr Anderson and Mr Wilson's replies sought to promote a positive image of the mining industry in Papua New Guinea and to reassure the public that appropriate legal and technological mechanisms were in place to ensure that the environmental damage of the type described by Mr Moutu would not happen. The exchange took place in 2008, shortly after Xstrata Copper took control of the Frieda River Project, bringing it into a full feasibility study. The takeover, completed in 2007, was received with optimism by the industry, the government of Papua New Guinea and the two provincial governments, who hoped that, once operational, the mine would provide much needed income, employment and services in what are some of the most remote areas in the country. The expectation was that, under Xstrata's management, the mine at Frieda River would go into full operation in 2012, with a projected life span of twenty-three years (Moutu 2008).

In a country with a rich and extensive history of hosting a variety of mining projects, planning for a mine at Frieda River proved to be a heated and highly contested issue. While the company embarked on a study aimed at assessing the technical and economic viability of the Project, the exchange of letters in *The National* and a broader debate that followed raised many issues and uncertainties regarding its social and ecological feasibility. The Project's location in the tributary of the Sepik River forced all parties involved to consider many potential perils and limitations associated with operating a large-scale, open-pit resource extraction project in that area. Topography of the terrain meant that the risks associated with its environmental impact were high, and the country's previous experience with large-scale foreign mining investments caused uncertainties about the feasibility of the claim that the mine, once operating, would deliver development to the people of Frieda and to the two provinces. Many of the social and environmental issues experienced around disastrous mines such as Panguna (May and Spriggs 1990), Ok Tedi (Banks and Ballard 1997; Kirsch 2006) and Porgera (Golub 2014) informed the ongoing debate over the Frieda River Project and were being negotiated in the pre-production, agreement-making phase of its development. The letter exchange in *The National* in 2008 reached a consensus that the long-term legacy of the early, pre-construction stages of planning for a mine at Frieda River would be crucial (Moutu 2008; Wilson 2008).

I arrived at Frieda in August 2011 – three years after the letter exchange in *The National* and forty-five years after the official recording of the area's rich gold and copper deposits paved the way for the resource extraction companies, attracted to the region by the promise of its great mineral wealth. By then it was clear that the mine's operations would not start in 2012, as global commodity prices meant that the company developing the Project decided to look into more economically viable infrastructural solutions for constructing a mine in this remote and humid corner of Papua New Guinea. The company's presence in the area was strong, however, and although its on-site exploration activities were largely limited to a series of sampling exercises, the Community Affairs department continued to actively engage with the Project's impact communities in the hope of securing and maintaining their consent for the Frieda River Project. I came to my field site on a company fixed-wing charter from Wewak. For someone with no knowledge of the terrain and with no local connections, this was and still remains the only way to get there. Ahead of my arrival, members of the company's Community Affairs team sent out a message through a local CB radio network to Paupe, Ok Isai and Wabia. The message said that a female researcher from the UK will be coming to Frieda and was hoping to stay in one of the local villages and learn about local communities and custom. When I disembarked at the company's Airstrip Camp, and once I had a chance to freshen up and eat something in the camp's canteen, I was told that one of the local leaders, pastor Jacob Aiyapanai, had invited me to spend a few nights at his house in Paupe with his daughter, Dinah, and her baby girl, Mawi. 'When the message came, we were in the garden house in Seligeme' – Dinah told me a few months after I first walked into the house in Paupe. 'That day dad came into the house and told me that a white girl will be coming to the village and that I am to go back to Paupe to look after her. So I took Mawi and we came back. We didn't know what to expect and I was nervous.' I too was nervous. My knowledge of Tok Pisin was limited to the basic words of greeting, and I did not know what to expect from the next few days.

I walked to Paupe on the evening of my second day at Frieda. With my field bag on my back, I tried not to slip as I navigated the narrow path leading from the airstrip to the village and Dinah's house. I was accompanied by Rondah – Dinah's aunt, who lived in a house nearby. She worked as a laundry assistant in the company camp and was heading back home at the end of her shift. Dinah and Mawi were waiting for me. We sat down on a wooden floor in a corridor of Dinah's house[11] – me against one wall, them against the other. I did not know then that I would spend almost a year sharing my life with Dinah, Mawi and the Paupe community. Over time, I came to think of her house nested in the southernmost corner of Paupe as home. Dinah's father's role as a local pastor and a spokesperson for the Paiyamo meant that I had easy access to community meetings, which often took place in a clearing between Dinah's house and kitchen building, right outside my bedroom window. As I became fluent in Tok Pisin, I could talk to the pastor and other members of the community more freely and was able to gradually explain to them my interest in the Frieda area and the Frieda River Project. I learned a great deal from the people of Paupe and owe them a debt of gratitude for the care they extended to me and the patience with which they helped me understand their world and their role as an impact community of the Frieda River Project.

A Note on Methodology

I was drawn to Frieda by my personal as well as professional curiosity. With a background in sustainable development and social anthropology, and a strong interest in social and environmental business accountability, I was intrigued by the ways in which different people and cultures came together, made sense of each other and tried to negotiate shared terms of reference to then use those terms in a process of planning for and managing a large-scale resource extraction project. I was by no means the first or the only anthropologist drawn to this world of ontological juxtapositions. In Papua New Guinea alone, many researchers have turned their analytical gazes to this interplay between indigenous cosmologies and corporate and governmental worlds. They took advantage of their associations with people with whom they worked to offer great and informative insights into indigenous perspectives on mining. Unsurprisingly, considering the nature of the industry, many of those analytical engagements focused on indigenous ways of experiencing land, presenting it as constitutive of personhoods and in doing so challenging a whole array of the industry's ideas about land and its resources as tradable market commodities.[12] A large number of ethnographic accounts put to the test the rigid forms and prescriptions associated with the ideology of land ownership, which provides foundations for many indigenous groups' involvement with the mining industry,[13] while others offer insights into the role of kinship and knowledge practices in resource extraction contexts.[14]

In the last decade, anthropology has turned considerable attention to the resource projects' state and corporate stakeholders. Companies and governments emerged from this exercise in anthropological scrutiny as products of relations that created them, animated by a complex and ongoing negotiation of roles and responsibilities, underpinned by the asymmetry of power and influence as demonstrated, for example, in Stuart Kirsch's *Mining Capitalism* (2014a). Alex Golub (2014) brought back to life the biblical leviathan and used it to investigate the turbulent and opportunistic relationship between the Porgera gold mine in the country's Enga Province ('the mine'), 'the state' of PNG and 'the Ipili', Porgera mine's recognised landowners. This analytical approach allowed him to begin to unravel the seemingly rigid boundaries between the different stakeholder groups and question the constitution of different stakeholder categories (see chapter 1). Elsewhere, two detailed ethnographic studies, produced in the last decade offer valuable insights into the inner workings of mining corporations: Welker (2014) drew on her extensive field research at the American headquarters of a mining corporation and its mining project in Indonesia to analyse ways in which the company was constituted and enacted in and beyond the mining locality; while Rajak (2011) used the findings of her research inside a mining corporation to reflect on the global corporate social responsibility movement. Both authors painted a picture of mining corporations as multilayered, multifaceted entities, with Welker describing them as 'inherently unstable and indeterminate, multiply authored, always in flux and comprising of material and immaterial parts' (2014: 4). By approaching the corporation as enacted, Welker was able 'to bring together the "idea" of the corporation as an actor endowed with particular goals and rationalities with the corresponding "system" of material relations and practices' (ibid.).

The results of these studies reaffirmed the positioning of mining projects in a chain of increasingly anonymous interactions ultimately reaching all the way through to stock markets and market-oriented shareholders, and it is in the context of these broad, contemporary spheres of interrelations and influence that the ethnographic material presented in this book must be read. However, the book itself focuses on processes taking place on the ground, in the Project location – away from company boardrooms and consumer markets – in a place largely hidden from view by the mountainous terrain and a thick rainforest overlay. The book is based on research conducted in the Project area over a period of eleven months in 2011 and 2012 and a follow up visit in February and March 2018. The research activities consisted of participant observation, informal conversations, focus groups as well as semi-structured interviews. To capture the diversity of perspectives and the meanings and histories of 'a mine' at the Frieda River, findings of a traditional village-based ethnography in Paupe were supplemented with data gathered through research engagements with Xstrata's on-site community affairs personnel as well as individuals representing national and provincial governments.

I was in touch with Xstrata before I arrived in Papua New Guinea. Although the company agreed to offer me transport assistance and to host me at the Airstrip Camp for a couple of nights at the start of my fieldwork, I got a distinct impression that they did not feel comfortable with my interest in the Project. In all early communication, they were careful not to commit to participating in my research. There is no natural position for anthropology in the contested field of resource relations.[15] Back in 2003, Ballard and Banks (2003: 290) spoke of the 'enduring opacity of mining corporations' and their 'notorious reluctance to expose themselves directly to ethnographic scrutiny'. The ideology of corporate social responsibility, which continued to increase in influence in the global industry in the first decade of the twenty-first century meant that by the time Kemp and Owen (2013: 534) revisited this claim ten years later it was clear that:

> many social scientists have for some time had direct access to the internal domain of the mining corporation through consulting, research and auditing/assurance work. Today the suggestion that companies are reluctant to expose themselves to external scrutiny is no longer valid in the way described by Banks and Ballard although caveats may still apply in terms of the release and disclosure of findings and results.

Formally, I never had any affiliation with Xstrata and maintained my position as an independent researcher throughout the duration of this study. Once at Frieda, I was able to develop positive personal relations with the company's ground personnel. The process took time, but eventually I was able to secure the management's permission to conduct research among the company's on-site employees, in particular – members of the Community Affairs team, whose role by definition positioned them at a difficult intersection between company structures and procedures and interpersonal relations they developed with members of the local communities. With the team's help, I was able to further expand the scope of my research and interview members of the provincial and national governments, who represented state interests at the negotiating tables at Frieda. These research encounters allowed me to develop a much deeper understanding of the process of planning for large-scale mineral development at Frieda and introduced auxiliary detail and nuance into the way in which I began to perceive and understand local communities' strategies in relation to the Frieda River Project. In the chapters that follow, I draw from the material collected among the company personnel to offer insights into the ways in which the company was enacted in the Project location and demonstrate how conflicting epistemologies came together in a highly contested network of social relations.

Throughout my time in the field, I was very aware of the sensitivities associated with my chosen research position and my decision to involve company personnel and government representatives in my work. I was alert to the power inequalities at play and the fact that the different actors involved in the process of planning for a mine at Frieda often held conflicting interests and made differing demands of the Project and its other stakeholders. Although my core research focus was the Paiyamo, my approach was driven by the recognition of the divergent world views and perspectives at play in the process of engagement between the company and local communities. I cannot claim here that I remained a fully neutral observer throughout my fieldwork – although my dual training in social anthropology and sustainable development helped me greatly to be empathetic towards people representing different positions at the Project and to see the same situation from a number of perspectives. In my work, I was guided by the principles akin to what Thomas Kirsch (2007: 98) described in terms of 'principled symmetry', which 'stresses the researcher's moral responsibility and sociopolitical accountability' towards all interlocutors. I strived at all times to maintain personal as well as professional integrity – that is, earn and return the trust I was given and to tell to the best of my ability the story of the Paiyamo and their history of the Frieda River Project.

Chapter Outline

This book tells a story of the Frieda River Project from the perspective of the Paiyamo – one of the Project's impact communities. Retracing local histories of the Project, the book confronts a set of assumptions about the constitution and appearance of 'a mine' to highlight and unpack some of the inherent environmental, social and governance complexities of the Frieda River Project. It builds on local ideas about knowledge, relations and ways of affecting change to explore ways in which the Paiyamo came to experience and understand the mine at the Frieda River and the strategies they used to engage with other stakeholders of the Frieda River Project and work towards that vision. To that end, the book offers an insight into contested processes of preparing for a large-scale resource extraction project as they were actually taking place, in the Project locality. Demonstrating the difference that perspective makes, the book explores what a mine would look like if we took something other than its physical industrial form as a starting point of analysis. This question is of particular importance for resource exploration projects such as the Frieda River Project, where the mine does not yet have a physical form but – as I aim to show here – from the perspective of the Paiyamo is already happening.

Chapter 1, 'Role Play', begins to unpack the history of the Frieda River Project. Following the timeline drawn from the Project's official documentation, it describes the first encounters between the Paiyamo and resource exploration companies, dating back to 1969 when the Carpentaria Exploration

Company began exploring the area's copper and gold mineralisation with the aim of assessing its feasibility for a large-scale mining development. A year later, construction of the Frieda River Airstrip began, exposing the Paiyamo to divergent ways of thinking about land as a tradeable commodity. Consequences of early misunderstandings between the company, the government and the community continue to unravel and inform the processes of agreement-making at the project and will be discussed in more detail in chapter 4. Those initial encounters formed the basis of what has since become known as stakeholder engagement activities at the Frieda River Project. The chapter introduces key actors in those activities, grouped under the headings of the company, the government and the impact communities, and raises important questions of ownership, power and control over the future of the Project. Using Strathern's theorising about categorical and interpersonal relations, it problematises stakeholder categories, presenting stakeholders as products of complex and dynamic relations, emergent from an intricate interplay between prescribed roles, expectations and relations through which those roles and categories are brought to life and awarded consequence.

The title of the second chapter, 'Making up the Paiyamo' is borrowed from Ian Hacking (1986) and his description of a process through which the nineteenth-century preoccupation with the proliferation of labels in the language of social control enabled the process of making up what he referred to as new 'kinds of people'. In a version of this process, and in the context of the Frieda River Project, the Paiyamo came to realise that in order to assert a claim to benefits from the Project they were expected to become of 'a kind' and become known to the company as the Project's landowners. They were not passively altered into landowners but actively transformed themselves in the hope of affecting a range of current and future opportunities. Continuing to unpack the history of the Project, the chapter shifts perspective to focus on Paiyamo narratives and the stories they tell about the ways they were, the ways they are and what kinds of processes led from one state of being to another. It describes how groups of Paiyamo speakers, dispersed over large stretches of land on the banks of the Frieda River, came together to form a single community in Paupe. The story serves here as an entry point into an exploration of the Paiyamo past, reaching back to the days of ancestors, mythical beings, magic and *kastom*[16] and their first encounters with colonial administration and resource exploration companies. Focusing on local experiences of time, the chapter challenges ways in which anthropology has theorised about temporalities in Melanesia in terms of epochs separated by dramatic, near complete transformation set in opposition to evolutionary ideas about progress, advancement and development. It shows that in local understandings of time, the present and future complete the past, and the meaning of the past can be altered to direct present and future events and opportunities. The chapter uses Paiyamo narratives to describe how the Frieda River Project opened a discursive space that revealed new capacities, meanings and power in events and stories of old. For the Paiyamo, no event is immediately knowable; its true meaning and effect can only ever become known over time – like when they told me that the story of Afek is an 'old story' but that now things are happening, and the story is 'becoming true'. This is critical to understanding divergent perspectives on 'a mine' at Frieda River. While at the time of my research the company and the government approached the mine as a future project, for the Paiyamo, the mine began in the time of Afek and the ancestors and it was already happening – it had just not yet revealed itself.

Chapter 3, 'Make-Believe', describes ways in which the Paiyamo engaged with supernatural forces within their world. It follows Lohmann (2003: 183) in modelling the supernatural as 'a volition that makes things happen, but is not always visible' and explains how this etic category could be used to describe *masalai* (land spirits), the Christian God as well as the mine at Frieda River. In doing so, the chapter reveals a form of engagement, made possible by Paiyamo ideas about personhoods, knowledge and social relations as tools for transformation. Exploring the religious lives of the Paiyamo, it describes ways in which they enacted a series of beliefs in the hope of making them happen. Addressing the tensions in anthropological debates about dividual and individual models of personhood, the chapter follows LiPuma (1998: 61) in focusing on uncovering the conditions 'under which dividual and individual aspects of personhood emerge and are hidden'. It describes elements of traditional religious doctrine, based on a welfare cult performed by men, to expose a complex and deeply relational system of careful management and strategic disclosure of important knowledge about the world, embedded in the local aesthetic of revelation – which plays an important part in the way in which the Paiyamo thought about and got to know the Frieda River Project. The chapter continues with the story of the arrival of Christianity in Paupe. Christianity held a particular appeal for the Paiyamo, granting them membership in a transnational and transracial system and allowing them, to paraphrase Robbins (1998b), to partake in a future promise extending beyond their current capacities. Elaborating on the theme of the importance of perspective, the chapter describes how while I struggled to understand the interplay between what I saw as two distinctive religious domains, the Paupe community experienced their religious lives in a way in which new religious elements dramatically expanded the limits of what could be believed and, therefore, of what might happen.

Chapter 4 exposes the changing politics of land and group affiliation in the context of commodification of land, which, although not fully realised, is already occurring, driven by anticipation of the mine at Frieda River. Continuing to explore the themes of knowledge and making the knowledge effective, the chapter uses a metanarrative of a document called 'The Book of Paiyamo' to discuss ways in which the encounters with the company developing the Project exposed the Paiyamo to new ways of 'doing knowledge' and making it visible. The chapter shows how, in response to the company and, to an extent, the government's demand for specific kinds of evidence, the Paiyamo embarked on a project of making their knowledge of relationships with the land, and with each other, appear in a particular form. The chapter pulls together themes of knowledge and its forms and the role they play in the politics surrounding the Frieda River Project. It uses 'The Book of Paiyamo' as an entry point to a broader analysis of Paiyamo ideas of knowledge and personhoods, which judge knowledge and relations against the transformative effects these have on the world. The chapter provides insights into the ways in which the Paupe community interpreted and responded to the company's epistemologies and developed new strategies in an effort to become known to the Xstrata and compete for recognition with the Project's other impact communities. It further problematises labels used by the company and the government in an effort to get to know the Project's local communities, and signals issues that have already emerged as a result of historical attempts to fit traditionally open and changeable social structures into a rigid category of 'landowning clans'.

The changing politics of land and new ways of 'doing knowledge' and evidencing claims are just some of the ways in which, seen from the perspective of the Paiyamo, the mine at Frieda has already started showing effect. Chapter 5, 'Precipitating the Mine', further unpacks the ways in which the Paiyamo experience the effects of the mine, which have already started to show in people and places. Although they voiced concerns about some negative changes they have noticed in their community, they continued to talk about the mine at Frieda in aspirational terms, describing the Project as a pathway to better livelihoods and development, and the levels of support for the Project in Paupe remained high. The chapter discusses how, in the absence of government services, civic infrastructure and access to education and employment opportunities, the Paiyamo looked to the company for what traditionally would have been considered responsibilities of the state. Informed by over forty years of interactions with successive resource companies, they held high expectations of the kinds of support and benefits the company should provide and formed relationships with Xstrata's community relations and development personnel to maximise the chances of bringing those benefits about. The chapter talks about how, with the government largely absent and the company's focus on finances and preference for rigid knowledge practices, the Community Affairs officers at the Frieda

River Project were left to bridge the relational gap between the company and the increasingly impatient and disillusioned local communities. This discussion is embedded in discourses of corporate social responsibility (CSR) and sustainable development (SD) and how these were interpreted and enacted by the company operating the Project. The chapter describes ways in which the company used the rhetoric of CSR and SD to at once acknowledge and limit its local obligations and development assistance, made possible by its understanding of 'a mine' and insistence that, while it may be there one day, there is currently no mine at Frieda River.

I am not in a position to judge whether the company was unable or unwilling to recognise that the Project's local stakeholders in Paupe had their own vision of what constitutes 'a mine' and that they spoke to that vision during stakeholder engagement activities, adding further complexity to an already difficult mineral development venture. As the material presented in this book shows, this vision was deeply embedded in local cosmology and the Paiyamo ideas of knowledge, transformation and personhoods, which shaped how they saw, experienced and interpreted the Frieda River Project. Through eliciting those ideas and focusing on the perspective of the Paiyamo, this book seeks to create a discursive space in which, despite the absence of the mining pit and industrial mineral extraction infrastructure, the mine at Frieda River can be brought to life and made visible.

Notes

1. EIS is an evaluation of the project's feasibility in its natural and social environment settings, based on comprehensive social and environmental scoping studies. All resource extraction companies operating in the country are legally required to produce an EIS for a proposed project and submit it to PNG authorities for approval as part of the permitting process prior to commencing construction.
2. In 2019, PanAust submitted an EIS for the Frieda River Project to the government. At the time of writing, the Project has not yet been developed, and its future remains uncertain.
3. Telefols is a local word used for Telefomin, a language group inhabiting areas to the north-west of Paupe (see Jorgensen 1981). Two Telefomin communities, Ok Isai and Wabia, are among groups formally recognised as landowners in the Frieda River Project area. In the Telefomin language, the word *bangri* was used to describe a stone adze, which used to serve as a popular tool in the area.
4. In 2011 and 2012 the story of Afek and *bangri* was commonly told in Paupe. When I returned to the community for a period of supplementary fieldwork in 2018, the relationship between the Paiyamo and the Telefomin communities was strained, as conflict emerged over a land boundary between the Telefomin

and the Melefale during a supplementary land boundary demarcation exercise in the Project area. Some of the Paiyamo, pastor Jacob included, no longer claimed that the Telefol story of *bangri* was also their story and placed more emphasis on other narratives asserting their own rights over land in the Project area.

5. The mythical ancestress of Telefomin (see Craig and Hyndman 1990).
6. String bags (see MacKenzie 1991).
7. Three versions of this story were previously recorded in the Frieda area by Dan Jorgensen. See Jorgensen (1997).
8. See also West's (2006) work on how Gimi speakers in the Eastern Highlands of PNG imagined their future and development in light of mining prospects in their area.
9. Aside from a couple of field reports written by Dan Jorgensen, based on his short visit to Paupe in 1984, there were to date no detailed ethnographic accounts of this greatly dynamic and innovative community that has over the years become a stakeholder to one of the biggest prospective mining projects in the Pacific region. This book fills an important gap in the anthropological topography of this part of the world, placing the Paiyamo between the well-researched Telefomin communities to the south-east (Jorgensen 1981, 1990a, 1990b, 1997, 2006, 2014) and the Miyanmin groups to the west (Morren 1979, 1986, 1981; Gardner 1980, 1983, 1987, 2004), Duranmin to the south (Lohmann 2000, 2001, 2003, 2007) and further on the fringes of what anthropologists came to know as the Min region of Papua New Guinea (Barth 1975, 1987, 1990; Brumbaugh 1990; Craig and Hyndman 1990; MacKenzie 1991; Polier 1992; Robbins 1998a, 1998b, 2001, 2007; Crook 1999, 2004a, 2004b, 2007a, 2007b; Brutti 2000, 2001, 2007 among many others). In this book, I do not seek to establish whether or not the Paiyamo are 'Min'. However, ethnographic accounts from the Min region have guided me through my initial engagement with the cultural interface that is the Frieda River Project.
10. For more information about Panguna, see May and Spriggs (1990); Connell (1991); Filer (1997b); Filer and Macintyre (2006) and Filer and Imbun (2009). See also Banks and Ballard (1997) and Kirsch (2007) for more detail about Ok Tedi.
11. Dinah's household consisted of a *haus slip*, a raised structure with weathered timber flooring and a tin roof with doors leading to three bedrooms from the central corridor; and a *haus kuk* – a kitchen, built out of bush materials, with an open fire pit where all meals were prepared and consumed.
12. For example, Weiner (1991, 1998b); Strathern and Stewart (1998a, 1998b); Rumsey and Weiner (2001); Leach (2003, 2014); Kirsch (2004); Strathern (2004); Anderson (2011); Stead (2017).
13. For example, Jorgensen (1997, 2007a); Power (1997); Kalinoe (2004); Leach (2004); Rumsey and Weiner (2004); Filer (2006).

14. For example, Crook (2007a); Gilberthorpe (2007); Macintyre and Foale (2013).
15. One of the best-known exchanges on the positionality of anthropologists on resource projects in PNG comes from Stuart Kirsch and Colin Filer. Kirsch (1996: 15) argued that anthropologists should actively advocate for local communities in resource conflicts even if this 'may compromise our status as impartial social scientists' (see also Kirsch 2002). Filer (1996: 26), on the other hand, held a strong belief that 'anthropologists may do more for the indigenous people of Papua New Guinea in their capacity as honest brokers in the negotiation of development or compensation deals between mining companies and local communities, than in their capacity as advocates for the anti-corporate alliance'.
16. *Kastom* (from the English 'custom') is a highly politicised concept in Melanesia (see chapter 3). This book follows the Paiyamo in their use of the term *kastom* to explain the local ways of world-making and being in the world and a range of contemporary practices 'which draw on tradition insofar as tradition is part of the present' (Strathern 2004: 2).

Illustration 1.1. A boy from Paupe watching the company's activities at the airstrip. Photograph by the author.

1

Role Play

Introduction

In 1870, a German physiologist, Lumbar Hermann, published an image of a white grid on a black background. As people moved their eyes around the image, multiple black dots seemed to appear and disappear at the grid's intersections. However, as they tried to focus on just one particular dot, it would vanish. The optical illusion meant that the more they focused on just one point, the less visible it became. For the large part, this is what trying to pin down and define stakeholder categories in resource contexts feels like. Looking at the intersecting grid of relations and dependencies, albeit far from a symmetrical one, we are constantly distracted by glimpses of activity as categories appear and disappear quickly in different parts of the image. However, if we only scrutinise one of those intersections and attempt to study it in isolation the category we are trying to isolate vanishes and what we are left with is constant and relentless movement in our peripheral vision. This is because stakeholder categories do not exist in isolation. They are reflections of a series of ideas about rights and responsibilities, animated by a dynamic matrix of social, legal, economic and technological mechanisms and relations.

This chapter uses narratives of the first encounters between the Paiyamo and resource exploration companies to begin to unpack the history of the Frieda River Project and introduce some of the interest groups the Project has brought into being. Those groups are commonly known as the Project's key stakeholders, and in very general terms can be grouped under three headings – the company, the government and the impact communities. In the mid-1980s, Freeman's stakeholder theory offered a new approach to the study and practice of organisational behaviour and business. It recognised corporate stakeholders, defining them as 'groups and individuals who can affect, or are affected by, the achievement of an organisation's missions' (Freeman, 2010: 52),[1] and placed them firmly on the spectrum of managerial and ethical concerns.[2] Over the years, the definition of a stakeholder has evolved. But far from becoming increasingly normative, it has progressed towards a growing recognition of the ambiguity of stakeholder roles and the need to analyse them in the context of the social relations and systems that produced them. In this process, stakeholders have been conceptualised less in terms of predefined actors and positions and more as arbitrary concepts that exist only insofar as people can agree on their goals, boundaries and usefulness (Ramirez 1999; Wolfe and Putler 2002).[3] This chapter problematises the notion of stakeholders and raises important questions about power, ownership and control over the future of the Frieda River Project. It presents the Project's stakeholders as products of complex and dynamic relations, emergent from an intricate interplay between prescribed roles, expectations and relations through which stakeholder categories are brought to life and awarded consequence.

Stakeholder Modelling and the PNG Mining Industry

Much of the thinking about the roles and responsibilities at resource extraction projects is framed around the notion of stakeholder categories. These categories group people of a given type under a particular label and often award them with prescriptive roles and responsibilities. Like in many countries around the world, in Papua New Guinea, national legislation makes the state the owner of the country's mineral resources and the regulator of the extractive sector. In order to capitalise on these resources, the government enters into agreements with mining companies, which have the financial and technological capacity to operate resource extraction projects. The country's first large-scale mine, the Panguna mine on the island of Bougainville, began its operations in 1972. The Ok Tedi mine in the country's Western Province followed twelve years later. In the 1980s and 1990s respectively, the two mines were sites of dramatic events that exposed the limitations of what can be roughly described as an early stakeholder model, which was predominantly based on agreements between the state and companies. At the Panguna mine, exclusion of local people from decision-making processes combined with the weak relationship between the government and the very people it was meant to serve, led to a breakdown of social relations resulting in a civil conflict that claimed thousands of lives, the consequences of which are still felt in many parts of the country (Filer 1990; May and Spriggs 1990; Connell 1991). At the Ok Tedi mine, misunderstandings and negligence of local ontologies and contexts exacerbated environmental issues, eventually leading to a legal challenge against the company operating the mine and its subsequent withdrawal from the project (Banks and Ballard 1997; Banks 2002). Over time, the mine, which started off as a symbol of national optimism of the newly independent state of Papua New Guinea (King 1997), became one of the most infamous mining projects in the world and what Kirsch (2014a: 133) branded 'a slow motion environmental disaster'.

Coinciding with the growing recognition of the rights of the indigenous people worldwide in the 1980s and 1990s, the experience of the Panguna and Ok Tedi mines sent shockwaves through the sector. They have triggered a systemic response from mining investors, operators and regulators globally and have been linked to the industry's engagement with the discourse of social accountability and corporate social responsibility (Filer et al. 2008; see also chapter 5). In Papua New Guinea, they introduced new kinds of uncertainties into the mining industry (Golub 2014) and forced the state as well as corporate actors to acknowledge that an engagement model based largely on binary relations between the government and the mining company was neither suitable nor sustainable for large-scale resource extraction projects – also in economic and political terms. This realisation led to the expansion of the engagement model to include groups living in mining localities and affected by mining activities – often termed 'the landowners' (Imbun 2007).

From an administrative perspective, some 97 per cent of Papua New Guinea's land remains in customary ownership (Filer 1997b). In legal terms, however, this has firm limitations. As prescribed by the country's laws, although the land belongs to the people, the resources that lie underneath it belong to the state (Mining Act 1992). This legal set-up plays a crucial role in the context of resource extraction projects across Papua New Guinea, where the presence of industry interests serves as a catalyst for ascribing new roles and characteristics to the communities in mining localities as well as to the state. The category of landowner is often considered the most opportunistic and at the same time the most problematic of the three general categories making up the triad stakeholder model (Ballard and Banks 2003), not in the least due to the contested nature and meaning of land ownership in Papua New Guinea.[4] The emergence of the landowner label has been linked to the process of colonial codification, aimed at turning a fluid and flexible history of social organisation and territorial occupation into a legally recognised system of units and structures (Jorgensen 2007a; see also Allen 2013 and Harrison 1999), but it has since become a powerful means of engaging with state and corporate forces with the aim of capturing benefits from large-scale mining and petroleum projects (Filer 2006, 2011; Jorgensen 2006). The landowners (referred to also as local- host- or, as is the case for the Frieda River Project, impact communities) emerged as an important if ambiguous political and legal category, caught up in structural power inequalities that continue to characterise resource relations (Kirsch 2007).

Being able to claim membership of this category became, for many communities, a means through which they could assert a stake in a mining project, gain access to agreement-making processes and claim a share of the project's benefits. What followed was a process described by Ernst (2001) as an exercise in codification of collective life, introduced by joined corporate and state forces and embraced in one form or another by the communities themselves who, in many instances, spotted and skilfully utilised the great transformative potential embedded in their ability to act out a particular and peculiar ideology of 'landownership' (Filer 1997a: 156). Ernst (1999: 88) saw this process as 'making "entities" or things from what have been contingent categories' and branded it 'entification of social relations'.[5] Over time, it became apparent, however, that the fluid and flexible Papua New Guinean societies do not lend themselves easily to the fixed and rigid legal vision of an ancestral, landowning clan, as the category, designed to 'find', record and reflect, in reality created new types of relationships, opportunities and obligations – as the following chapters will show.[6]

The triad model, which recognises the state, the company and local communities as the three key stakeholder groups, continues to serve as a 'provisional analytical device' (Ballard and Banks 2003: 289) in the industry as well as academia, amid mounting acknowledgement that 'it has not generally served to capture much of the complexity of the relationships that form around mining as a site' (ibid.). As the sense of a mining community broadened, the significance of additional stakeholders such as NGOs, the media, legal agencies and global financial institutions rapidly expanded, forming what Ballard and Banks (2003: 304) glossed as 'the fourth estate' – included in resource relations 'by virtue of connections to or alliances with (rather than membership of) one of the three principal stakeholder categories'. More recently, Filer and Le Meur proposed extending the triad model to include a fourth element, which they call 'society'. They also reframe the original three stakeholder categories as spaces rather than conceptual entities and argue that:

> There is no large-scale mining project that involves just one company, one government agency or even one mine-affected community. Instead, there is a corporate space, a government space and a community space, each of which is liable to be occupied by a number of organisations or agencies in different relationships with each other, some of which count as political relationships, and within each of these organisations or agencies there are more political relationships. (2017: 22)

This view is reflective of a growing recognition in the anthropology of mining that categories, if held too tightly, can hinder rather than aid our understanding of complexities of stakeholder engagement at resource extraction projects. The discipline has turned much of its focus away from Scott's (1998) legibility apparatus – which awarded prescribed categories and labels with particular types of influence and particular types of consequentiality[7] – and towards nuanced processes and enactments of categories as they are taking place, in project locations (e.g. Golub 2014; Welker 2014). Inevitably, the closer we get to the ambiguities and complexities of relations on the ground, the more distant and devoid the categorical prescriptions become. Yet, it would be unwise to dismiss categorical prescriptions from any critical analysis. Neglecting them altogether is bound to introduce new kinds of misunderstandings and, ultimately, limit our capacity to observe and comprehend the complex social phenomena such as planning for large-scale mining developments. Here, I focus instead on the contested spaces in between stakeholder categories and their enactments in the project location. Categorical models tend to make assumptions about what the latter ought to look like. They often contain specific instructions to facilitate types of interactions and relations (e.g. through legal frameworks, employment contracts and bureaucratic procedures). But, in a version of a looping effect described by Mary Douglas (1986: 108), categorical models are themselves influenced by the way they are experienced and imagined in resource project localities. In her theorising about institutional reclassifying 'the people are tempted out of niches by new possibilities of exercising or evading control. Then they make new kinds of institutions, and the institutions make new labels, and the label makes new kinds of people'. A version of this process has led to the establishment of Paupe, an impact community of the Frieda River Project – as the next chapter will show.

'The Company Arrived in a Helicopter and Landed in Paupe'

The story of the industry's presence at Frieda goes back to 1969, when the Carpentaria Exploration Company (CEC), an exploration arm of an Australian company called Mount Isa Ltd, arrived at Frieda having obtained a Prospective Authority (EL-58) from the Papua New Guinean administration. I heard the Paiyamo describe the moment of CEC's arrival on a number of occasions. 'The company arrived in a helicopter and they landed in Paupe' – they would say. According to the story, the first arrivals were a team of geologists working for CEC, accompanied by a policeman and a kiap from Ambunti, a government station located on the Sepik River, north-east of Paupe. At the time, pastor Jacob was a very young boy, and he followed the others as they ran away to hide in the bush. 'It is because we were scared. We had never seen a helicopter before and we were worried that it was the army coming to attack us,' the pastor explained to me.[8] The rainforest, stretching along the riverbanks and further up the slopes beyond the horizon, plays an important part in many Paiyamo myths and stories. Full of uncertainty and largely impenetrable to an inexperienced navigator, it constitutes an integral part of life for the Paiyamo. Pastor Jacob's father, Aiyapanai, was one of the very few men who did not run away and was the first one to speak with the new arrivals. The policeman and the kiap disembarked from the helicopter and walked over to Aiyapanai. They asked him to identify the owner of the land on which they were standing. 'But Aiyapanai did not speak Tok Pisin and he did not understand the question so he just nodded. The kiap wrote down Aiyapanai's name. He thought Aiyapanai was the owner of the land and was saying yes to mining, giving permission for the company to come and work here.' Following this encounter, the visitors got to work. They travelled around the area, setting up temporary camps in the bush and collecting stone samples for testing at a junction between two main rivers – the Frieda River known locally as Neigeme, and Nena River, which the Paiyamo call Melegeme – in what is now the Project area. Every version of this story I heard during my stay in Paupe concluded with a comment about the time that had passed since these events, more often than not embedded in a biographical narrative of the teller (e.g. 'I was little when the company first came. Now I am an old and they are still here').

The stories of Paiyamo's first encounters with a mining company almost always combine feelings of fear and uncertainty with happiness at the new potential and prospects that the arrivals would bring into the area, which I will come back to in chapter 2. These stories came to constitute an important part of the Paiyamo mythology of the Frieda River Project, and I often heard members of the community saying that they did not understand back then what these events and new encounters meant but that their significance was revealed to them over time.

Building an Airstrip on the Bank of the Frieda River

CEC began work shortly after its first representatives arrived in the area in 1969. The little information I could find about their early activities at the Frieda River Project site suggested that the initial resource exploration continued for around three years and yielded some very promising initial results. Logistically, however, working in such a remote area proved challenging, especially considering the company's plans to expand and extend its activities on the ground. All company personnel had to be either flown in by a helicopter or face a long and tiring journey up the Sepik River and further south following the rapid currents of Frieda. The same applied to supplies, building materials and machinery. To enable plane transport into the area, preliminary plans were drawn up in 1970 to construct an airstrip on the western bank of the river, a short walk away from the southernmost corner of Paupe, on a strip of land known by the Paiyamo as Piamoku. The only available written account of this time comes from John Pasquarelli. In his personal notes, he recalls the early days of exploration at the Frieda River Project site and the task of constructing the airstrip in the middle of the bush on the steep shores of a fast flowing mountain river:

> [In 1970] Frank Martin told me that he had been approached by Carpentaria Exploration to build an airstrip on the banks of the Frieda River near the Paupe settlement and that McIntyres, a firm of engineers from Townsville, would be involved. From my briefings and my first expedition to the Frieda River in 1963 and subsequent visits, Frank was aware of the river, its rapids and shallows when the river was not flooding and he told me that a D6 dozer, International Drott tracked loader, grader and tip-trucks would have to be shipped to the site. Over a few beers we discussed the feasibility of constructing a jet-barge to be powered by Holden 308 V8 motors driving Hamilton Colorado jet units. Without further ado, Frank Martin planned and constructed a steel skinned jet-barge on the beach at Wewak. [He] took the barge from Wewak up the Sepik to the Middle Sepik port of Pagwi. The scene was set for the movement of machinery from Wewak to the Frieda River. Frank Martin had a good team working for him at Wewak Transport Service and a standout was Jimmy Gordon the ex-Public Works road master who came to the Frieda River to help with the construction of the airstrip. Frank told us that a D6 dozer, a Caterpillar 12 grader, an International Drott tracked loader, 4 Izuzu 6x6 tippers and steel trusses for an aircraft hangar would be the first loads to be shipped up the Frieda.
>
> The D6 dozer was the first load and it came out to Pagwi on a low loader and after lots of manoeuvring and swearing it was finally lashed down on the barge, followed by anxious checks as to how the barge was sitting in the water. It was a slow, hard grinding trip with a few scary moments but the D6 finally made it ashore to the Frieda River Airstrip site. A mishap such as a capsize could have been disastrous, indeed catastrophic if lives had been lost but the Sepik gods smiled on the venture and all equipment including some basic portable buildings made it safely to the airstrip site. The building of the Frieda River Airstrip is a monument to the drive and natural mechanical and engineering skills of Frank Martin and the hard work of his PNG workers, led by Jimmy Gordon. Norm Martin and myself did our bit as did Kinook, Sal, Weliwan, Bundi, Patoman, Asa and the Frieda and May River locals who were recruited as labourers. Helicopters played a crucial role in the development of the Frieda project and still do but contact with the wider world was achieved with the landing of Mount Isa Mine's Norman Britten Islander on the new Frieda River Airstrip.[9]

If Pasquarelli's timeline is correct, the airstrip construction at Piamoku was in full swing by 1972 when the land acquisition plans were drawn up. Located some 50 miles south from the mouth of the river, Piamoku was described in official documentation as some 20,466 hectares of what was officially recognised as land with 'native owners'. In documentation, it is mislabelled as 'Iamu' – a result of early miscommunication between the Paiyamo and the CEC. 'They asked us what this place was called. We told them we called it Piamoku. They must have heard us wrong and wrote down Iamu. That's the company for you,' a member of Paupe community told me once, shaking his head. For the purpose of land acquisition, Piamoku had to be carefully identified, measured and mapped. In this process, it was reduced to a prudently sketched-out image on a surveyor map, in which a series of numbers, lines and measurements aimed to turn an area of ancestral ground into a fixed, tradeable commodity (Figure 1.1.).

In line with the bureaucratic systems and attitudes of the time, the acquisition plans were accompanied by a note signed by Harry William West, a member of the colonial administration residing at the nearby May River station. According to the document, West was 'specifically charged under Section 85 of the Land Ordinance 1962 to establish, further or protect the interest of natives in or in relation to native land'. In the *Certificate of Alien Ability* dated 7 December 1971, West's signature attests to the fact that 'the native owners of the land are willing to sell the land to the Administration; the sale will not be detrimental to the best interests of the native owners or of their descendants either now or in the considerable future'. He goes on to declare that he had 'fully considered the question of reserving to the native owners and/or their descendants [the] right of hunting, gathering, col-

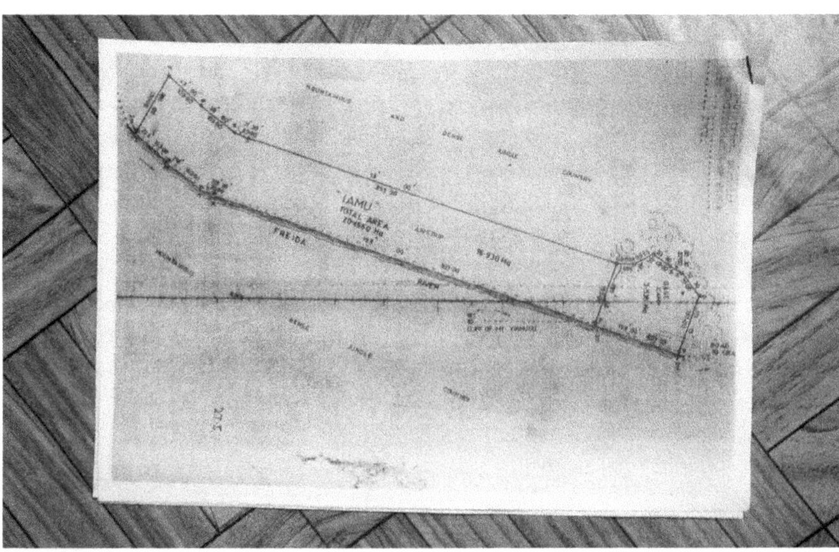

Figure 1.1. Piamoku ('Iamu') as presented in the sale of land documents in 1972. Photograph by the author.

lecting, fishing and access and recommended that no such reservations be made'. Six months later, in June 1972, another May River Post Patrol Officer, Dennis John Mamar, arrived at Frieda with a copy of the *Sale and transfer of land* contract. He was given the authority by the administration to execute the document, which was written in English and which outlined the terms and conditions of the sale. The contract was signed on 18 June 1972 by representatives of the colonial administration and the people living in Paupe, officially reassigning the ownership of Piamoku ('Iamu') to the administration, which then formally leased the land to the company.

Forty years later, in Paupe, pastor Jacob was now coming to me with an old paper folder from which he carefully took out a faded copy of the sale of land documentation.[10] He wanted to show it to me, he said, to help me see and understand Paupe's and the Paiyamo's role in the Frieda River Project. He handed the documents over and sat back against a wall on an uneven floor of Dinah's kitchen. The contract was drafted by representatives of the Papua New Guinean colonial administration. It branded the administration the purchaser of the land, and inhabitants of the Paupe village the land's vendors. In doing so, it attributed both parties with specific roles and powers for making the transaction possible and, from a corporate and administrative perspective at least, legally binding. In the formal signing of the contract, the vendors were represented by two men named in the document as Aiyapanai

Illustration 1.2. Frieda River Airstrip. Photograph by the author.

(pastor Jacob's father) and Nemasio. The two acted on behalf of twenty-five named members of the Paupe community listed in the 'schedule of owners' document attached to the contract.[11] The contract stated that:

> Said owners have agreed with the Purchaser to sell the said land and all the rights and interests . . . This deed witnesseth that in consideration of the sum of one hundred and twenty three dollars ($123.00) now paid by the Purchaser to the Vendors (the receipt of which sum the Vendors hereby acknowledge) the Vendors pursuant to the powers vested in them by virtue of native custom hereby sell transfer and convey to the Purchaser the said land and all right title and interest in therein of the said owners and of all other persons who otherwise would have subsequently become entitled to the said land by virtue of native custom in the same and every part thereof.

The document bore marks of a cross (x) by Aiyapanai and Nemasio's names, still clearly visible despite the passage of time. It was accompanied by Mamar's statement in which he certified that he had 'truly interpreted the contents of the said Transfer into the Pidgin language [Tok Pisin] understood by the Vendors'. He went on to state that he was 'certain that Vendors understood him', that they 'knew what land they were selling' and that they knew that 'they were selling the said land outright to the Purchaser'. But forty years later, the Paiyamo were challenging those claims. As I was told by one of the very few men in Paupe who knew English and could therefore read the original land acquisition documents:

> When in 1972 the government came to buy the land, there was an officer who spoke Tok Pisin to the white man. The District Administrator spoke Tok Pisin to the officer, who translated it to Iwam language. Then Tikap translated from Iwam to Paiyamo. Aiyapanai and the other man, they did not understand when they were told they were selling the land. They got a hundred and twenty Australian Dollars but they did not know Piamoku was being sold. So now we want the land back.[12]

On another occasion, I heard a different Paiyamo man say that Piamoku was his ancestral ground. That meant that it belonged to him as much as it ever did to Aiyapanai and Nemasio. The company may have bought the land from the two men in 1972, but they did not buy it from him and others living in Paupe during my stay there, which in his eyes made the company claim of ownership of the land invalid.

The story of the acquisition of Piamoku can be seen as an opening act in a role play the Paiyamo, the CEC and the colonial administration and then the Papua New Guinean government engaged in in the processes of planning for a mining development at Frieda. It highlights misunderstandings reaching far beyond linguistic miscomprehension, capturing an encounter between greatly varied cultural logics and divergent perspectives. In the eyes of the company and the colonial administration, a series of technical and bureaucratic exercises culminating in the contract of sale turned Piamoku into a fixed and tradable commodity for which a legal owner could be identified. This kind of process, common to resource extraction projects around the country, is based on a premise that there exists in Papua New Guinea an indigenous system of land ownership that can be written down by common law, turning a fluid history of indigenous land tenure into a system of discrete land units for which legal owners can be found (see Jorgensen 2007a).[13] But confronted with the relations among local people and their connection to the land, this premise loses its footing, exposing instead fundamental ontological misunderstandings at the Frieda River Project – as chapter 4 will show.

In the next part of this chapter, I introduce key actors actively engaged in the process of preparing for mining development at Frieda. For ease of presentation, I have divided the text into three subsections: the company, the government and impact communities. But far from endorsing the triad stakeholder model as an all-encompassing, universal mechanism for managing stakeholder relations at resource extraction projects, what I am hoping to show here and in the chapters that follow, with regards to the Frieda River Project, is the diversity of forms and activities that actually characterise such stakeholder engagement. Some of the discursive ground for the arguments I present here was laid out by Golub (2014) in his now well-read study of stakeholder relations at the Porgera mine in the Enga Province of Papua New Guinea. In his monograph, Golub described the company, the state as well as the landowners (the Ipili) as leviathans enacting stakeholder roles and in doing so bringing to life particular conceptual and institutional categories (such as 'the government'). Golub (2014: 75) offered an intricate ethnographic study of 'the way "the Ipili" and "the mine" slowly became assembled out of the personal networks of early explorers and entrepreneurs who lived and worked in the valley' and painted a picture of the mine as 'a complex sociotechnical system made up of a variety of human and nonhuman actors' that requires an 'extraordinary amount of coordination' (ibid.: 7). In his work, Golub set out to examine *who* 'the mine' is, embedded as it is in a complex and dynamic nexus of relations in the mining locality. In this book, I investigate *what* 'a mine' is. To an extent at least, this difference in emphasis can be explained by dramatically different Project realities on the ground at the Porgera mine and the Frieda River Project site. While Golub writes about a mine that began operations in 1990, the existence of which cannot be disputed, I am describing an advanced exploration project at a pre-construction and pre-operations stage of its life cycle where the existence of 'a mine', as I set out to show, is a matter of perspective.

The Company

I first arrived at Frieda in 2011, forty-five years after the area's deposits were officially recorded, and forty-two years after the first company began exploring Frieda's dense rainforest in search of its rich mineral assets. In that time, numerous companies have been formally involved in the project (see Table 1.1.) although from the perspective of people living in the mining locality this list is somewhat problematic. Numerous business mergers and takeovers meant that in many instances changes to company form and ownership did not demonstrate in quite the same way on the ground. In many cases, while the legal ownership changed hands, at least some of the ground personnel remained the same, putting on new uniforms and pledging new company allegiance.

At the time of research, the Frieda River Project was owned jointly by Xstrata Frieda River Ltd, a subsidiary of Xstrata Copper and part of the Xstrata Group (81.82%), and a Papua New Guinean incorporated company, Highlands Pacific Ltd (18.18%).[14] Although Highlands Pacific and Xstrata shared the ownership of the Project, the two companies had very distinctive roles. Highlands Pacific Ltd was the formal owner of the EL-58 and took part in strategic decision-making regarding the future of the mine. As a relatively small company with limited financial capital and technological capacities, Highlands Pacific Ltd relied on entering into partnerships with more powerful players such as Xstrata in order to advance the Project through the consecutive stages of its life cycle. Xstrata was the Project's manager. The company operated three semi-permanent camps in the area. There were also some temporary drilling sites scattered around the locations of Frieda's largest mineral deposits.

Even though I refer to it simply as 'the company' or Xstrata throughout this book, it is important to place the project operator in its broader organisational setting. Xstrata Frieda River Ltd was a Papua New Guinea registered company with corporate headquarters in the country's growing and busy capital – Port Moresby. It was embedded within a much larger, Australia-based parent company, Xstrata Copper, which, at the time, held a high profile portfolio of mines in Argentina, Australia, Canada, Chile, Peru, the Philippines and the United States of America. The Frieda River Project was its first and only business venture in Papua New Guinea. Both Xstrata Frieda River Ltd and Xstrata Copper belonged to a global diversified business group, Xstrata Plc. With headquarters in Zug, Switzerland, and a registered office in London, United Kingdom, Xstrata Plc was an Anglo-Swiss company that traded on both the London Stock Exchange and the SIX Swiss Exchange. It was first listed in 2002 and quickly grew from a minor corporate player to one of the big names in the global resource extraction industry. I

Table 1.1. Timeline of the Frieda River Project.

1966	Australian Bureau of Mineral Resources mapping exercise records gold and copper deposits at Frieda.
1968	PNG Administration issues an EL-58 to Mount Isa Mines Ltd and its subsidiary, the Carpentaria Exploration Company.
1969	Carpentaria Exploration Company begins exploration work at Frieda. By 1972, porphyry-style mineralisation is identified at the Koki and Horse deposits.
1973	Mount Isa Mines Ltd completes the first pre-feasibility study of the Frieda prospects.
1974	Mount Isa Mines Ltd enters into a joint venture agreement with a Japanese consortium led by Sumito Metal Mining Co. Ltd.
1974–78	Frieda prospects are managed by Frieda Copper Pty Ltd. The company is registered in PNG and wholly owned by Mount Isa Mines.
1978	Sumito Metal Mining Co. Ltd restructures into OMRD Frieda Co. Ltd.
1979	Two more companies join the Frieda venture. CRA Exploration Pty Ltd from Australia and Norddeutsche Affinerie from West Germany.
1987	Mount Isa Mines transfers its stake in the project to its wholly owned subsidiary, Highlands Gold Properties Pty Ltd.
1990	Mount Isa Mines sells Highlands Gold Properties Pty Ltd.
1993	CRA Exploration Pty Ltd sells the project interests to the operator of the joint venture, the Highlands Gold Properties Pty Ltd.
1994	Norddeutsche Affinerie follows in CRA's footsteps and sells the project interests to the operator of the joint venture, Highlands Gold Properties Pty Ltd.
1994–96	Highlands Gold Properties Pty Ltd completes exploration study of Nena, Koki and Horse/Ivaal deposits.
1996	Highlands Gold Properties Pty Ltd completes a feasibility study, including a landowner identification exercise.
1997	Placer Dome Inc. comes to an agreement with the Highlands Gold board to create a new company, Highlands Pacific, which purchases Highlands Gold Frieda assets.
1997	Highlands Pacific is floated on the Australian Stock Exchange. The company now owns 85% of the venture, with the remaining 15% in the ownership of the Japanese Consortium OMRD Frieda Co. Highlands Pacific begins searching for a new venture partner with major copper mine experience to progress the project through the feasibility study.
1998	Cyprus Climax Metals Company (a subsidiary of Cyprus Amax Minerals Company) enters the project as the third partner of the joint venture with Highlands Pacific Ltd and OMRD Frieda Co. Ltd.

(continued)

Table 1.1. continued

1999	After just one year, Cyprus Climax exits the project.
2002	Highlands Pacific enters into a joint venture agreement with Noranda Pacific and OMRD Frieda River Co.
2005–06	A series of takeovers introduces Xstrata into the joint venture, as Noranda Pacific merges with Falconbridge (2005), which is then acquired by Xstrata (2006).
2007	Xstrata Copper exercises a management option and Xstrata Frieda River Ltd enters into a joint venture with Highlands Pacific Ltd and OMRD Frieda River Co. Ltd to complete a scoping study.
2009	In the middle of an extensive scoping study, OMRD Frieda River Co. Ltd exits the Project.
2010	Xstrata Frieda River Ltd commences a full feasibility study of Frieda's deposits.
2011	Xstrata Frieda River Ltd completes an EIS for the Frieda River Project but does not submit it to the regulatory authorities as the Project is not considered financially feasible.
2012	Xstrata Frieda River Ltd extends the deadline for the feasibility study. The company advertises expression of interest for potential investors.
2013	Xstrata PLC merges with Glencore International PLC, forming Glencore Xstrata PLC. Xstrata Frieda River Ltd ceases to exist.
2014	PanAust Ltd acquires Glencore Xstrata PLC's interest in the Frieda River Project. Frieda River Limited is formed.
2015	A Chinese company, Guangdong Rising H.K. (Holding) Limited, acquires 100% ownership of PanAust Ltd.
2015–16	Frieda River Limited prepares an SML application and Proposals for Development (including a Feasibility Study) to satisfy a condition of the EL-58. These are submitted to and accepted by the MRA in June 2016. Frieda River Limited completes a scoping study, a pre-feasibility study and other studies to support the EIS, which it submits to Papua New Guinea's Conservation & Environment Protection Authority (CEPA) in December 2016.
2017	Key Project changes since the submission of the EIS to CEPA in December 2016 lead to the requirement for a new EIS and SIA.
2018	PanAust Ltd announces it has identified a new nation-building pathway for the Frieda River Project and reveals its plans for the 'Sepik Development Project'. The company completes a revised EIS for the Project and submits it to CEPA as part of the permitting process.
2019	The company lodges an amended Proposal for Development in relation to the Frieda River Project's Special Mining Lease application with the Mineral Resources Authority. Cobalt 27 Capital Corp. acquires Highlands Pacific Ltd and transfers the company's stake in the Frieda River Project to PanAust Ltd. PauAust Ltd becomes the sole owner and operator of the Project.

Source: The timeline is based on project documentation obtained from the MRA, and the Social Impact Assessment for the Project (Coffey 2018: 16–7).

am under no illusion as to the vastness of the business entity that was Xstrata or to the great complexity of global networks of interdependence that it helped to reproduce. But in my ethnographic work, I focused on how it was enacted on the ground in the mining locality. I spent a lot of time at Frieda, watching and talking to the company's on-site employees, whose formal work uniforms displayed the sharp corporate logo of Xstrata Copper next to softer lines depicting the meanders of the Frieda River, which became the company's emblem for the Frieda River Project.

With the exception of a small group of largely unskilled workers hired from the Project's impact communities, the company's workforce was flown into the area on a four-weeks on, two-weeks off roster. The company's fixed-wing charter made regular, scheduled trips all over the country, bringing to Frieda people and goods from Mt Hagen, Lae, Wewak, Port Moresby and beyond. Workers were accommodated in one of the Project's residential camps, with a large majority staying either at the Airstrip Camp, occupying the northern part of the contested site of Piamoku, or the larger Base Camp located further inland, accessible from the airstrip by a regular helicopter shuttle service. I did not conduct any research in the Base Camp, but I had relatively unrestricted access to the airstrip, where a large proportion of meetings between the company, the government representatives and the impact communities took place, providing fertile ground for conducting ethnographic observations.

In preparation for landing, the company's fixed-wing charter plane would approach the airstrip from the south – the noise of its propellers, trapped in the valley and carried forwards by the fast flowing waters of the Frieda River, constituted a regular feature of the area's soundscape. The flight schedule and frequency changed periodically but generally the plane's arrivals and departures served as an important marker of time both in the company camps and for the Project's impact communities. The plane was used to transport the company's employees and guests. Unoccupied seats and free hold space of the aircraft were filled with goods and supplies brought into the camps from Port Moresby and further afield. Although only a relatively short walk along a narrow path tracing the western bank of the Frieda River separated the airstrip from the southernmost corner of Paupe village, in many ways the two places were very far apart. The temporary village houses constructed from bush materials and dimly lit by fire pits and kerosene lamps contrasted sharply with the camp's relatively ordered and modern infrastructure. In 2011, the Airstrip Camp was home to Xstrata's Community Affairs and medical teams and accommodated a number of the company's guests, visiting experts and manual labourers. All of them stayed in air-conditioned, furnished rooms and had access to toilet and shower facilities. The rooms were serviced regularly, and a laundry service was provided on-site. Three times a day the camp's canteen served generous, Australian-style meals, and the smell of morning coffee, bacon and eggs would often reach me in Paupe, carried over by warm, southerly winds. A large petrol generator provided a steady supply of electricity to every on-site room and building, and in the evenings, the warm glow of camp lights was visible from two of Paupe's three corners. Crucially, phones and satellite internet available on-site facilitated communication within as well as outside the Project area which meant that the Airstrip Camp was, for company employees as well as members of the local communities, a valuable and lively hub of gossip and information.

A typical day at the Airstrip Camp started with a morning meeting after which members of staff dispersed to embark on their daily tasks. Working days in the camp were long, starting immediately after breakfast and ending with the canteen bell calling everyone in for the evening meal. In very general terms, the company's activities at the Frieda River Project were divided into two distinctive but partially overlapping spheres. The operations division, headed by the operations manager, was responsible for the daily running of the company's on-site activities. This included transportation of people and goods in and out of the Project area, facilitating technical experts' visits and work schedules and overseeing the facilities provided in the company camps as well as the activities at drilling sites scattered around the rainforest. The division employed a number of local community members in low and semi-skilled positions, with some employees receiving substantial training as part of their work agreements. The number of local employees working on-site at any given time fluctuated greatly, dependent as it was on the schedule and kind of work being carried out at any given time. Then there was the Community Affairs team. The team was established when the company took over the management of the Project in 2007. As described in its sustainability report for that year, the team was tasked with enacting the company's corporate social responsibility ethos and community relations policies – that is, fine tuning and delivery of the company's Community Development Plan and ensuring efficient and clear communication and good relations between the Project's local stakeholders (Xstrata 2008). The team's members, known commonly as CA officers or CAs, served as the first point of contact for members of the impact communities and, in many instances, government representatives. The team's office was located some distance away from the Airstrip Camp's main infrastructure, at the northern edge of Piamoku. Just behind the office, a wide gate and high fencing separated the company camp from its external environment. Members of impact communities were allowed to cross the gate but did not have the permission to go past the CA office and into the camp without invitation and/or appropriate authorisation.

The Community Affairs team fluctuated in size but generally consisted of about ten CA officers, led by three CA superintendents and headed by the CA manager. With the exception of the manager, who was based in Xstrata

Illustration 1.3. 'No entry' sign at the entrance gate to the Airstrip Camp. Photograph by the author.

Copper's office in Brisbane and who made regular if brief visits to the Project site, the remainder of the team were on-site full time. I spent a lot of time talking to different members of the team and observing them at work either in the camp or in one of the impact villages. The length, scope and level of formality of those encounters varied greatly. With the exception of two rounds of formal interviews I conducted with the members of the team, where I asked each of them a set of pre-prepared questions, these conversations were generally unstructured, sometimes trying, sometimes encouraging and almost always insightful, albeit at times in very surprising ways. The time I spent with the CA officers allowed me to see many of the events described here from their perspective, gain insights into ways in which they conceived of and performed their professional roles, and understand different aspects of the company's activities and internal politics at the Frieda River Project (see chapter 5).

The Government

While the company personnel's presence on-site was strong and their interactions with impact communities frequent, the government's involvement in the local nexus of relations was marked largely by its absence. Following independence in 1975, the government of Papua New Guinea turned to the discourse of development in search of national identity (Jorgensen 2007a). Exploration of the country's natural resources, driven by increasing global demand for minerals, revealed Papua New Guinea's mineral wealth and provided the newly established government with great economic opportunities. Analysed within development discourse, the mining industry seemed to fit perfectly with the country's aspirations – bearing a promise of bringing crucial income to the country's government and delivering infrastructure and services to its citizens (Gilberthorpe and Banks 2012). Over time, the government's anticipation of a mineral-resources led development became shared by communities across the country, including the impact communities at the Frieda River Project. In a relatively short period of time, however, Papua New Guinea started displaying signs that the country was affected by a 'resource curse' (Ballard and Banks 2003), as its future development became increasingly dependent on its ability to attract foreign investment in mining (Connell 2005). In consequence, the Papua New Guinean government has found itself in a very difficult and often conflicting position, having to balance its duties towards its people with the need to create an appealing and attractive operating environment for foreign investment in the growing resource extraction industry (Golub 2014).

At the Frieda River Project, the history of state involvement predates independence and begins in 1962, when a colonial patrol officer arrived in the area from Ambunti station in the Sepik. Early encounters between the Paiyamo and, at the time, the colonial administration were influenced by the administration's efforts to categorise the country's population and turn the scattered pattern of occupation along the banks of the Frieda River into a system of villages – fixed administrative units that could lend themselves to the state apparatus of governance and control (see chapter 2). A reflection of this process can be found in the 1972 sale of land documentation, which classified the people of Frieda according to the pattern of newly established administrative units and villages, using Paupe village as a fixed entity representing and, at the same time, legitimising people's rights to Piamoku. Over time, as chapter 2 will show, those early exercises and experiments with settlement patterns became experiments with social organisation and laid the groundwork for the emergence of Paupe as an impact community of the Frieda River Project.

Except for some myths and stories told to me by my hosts in Paupe, I have found little record of the beginnings of the relationship between the Paiyamo and the colonial administration (and, later, the government of Papua New Guinea). But the little information I did find in reports produced by colonial patrol officers suggests that since its early days this relationship was weak and limited to sporadic and brief encounters. To this day, the Paiyamo feel very little obligation towards the state and are more likely to look to the corporation for what would traditionally be seen as the responsibility of the government (such as infrastructure and service delivery). For all intents and purposes, the way the Paiyamo see it, they do not have a relationship with the government independently of the Frieda River Project. This experience of detachment from the state is not unique to the Paiyamo or the Project. It is common in many areas of the country, where conversations about the government's involvement with resource extraction projects tend to focus on the government's absence from mining localities rather than its presence (see chapter 5).

At the Frieda River Project, it was explained to me by a member of a government agency, the government was first and foremost the regulator. It developed and implemented the legal framework within which mining development takes place. This includes the Mining Act 1992, which contains principal regulations regarding mining activities in Papua New Guinea; the Land Act (1996, revised in 2010), which regulates ownership, access to and use of state-owned land and land held under customary tenure; and the Mining (Safety) Act (1977, revised in 2007), which includes regulatory provisions regarding safety of mining infrastructure and mining operations around the country. The Environmental Impact Statements for mining projects in the country are assessed under the Environment Act (2000), administered by the Conservation and Environment Protection Agency (CEPA). The Mining Act 1992 clearly identifies the state of Papua New Guinea as the sole owner of the country's mineral resources, making mining companies liable to pay royalties to the state (the equivalent of 2 per cent of the net proceeds from the sale of minerals). It also gives the state a guaranteed right to a participating interest of up to 30 per cent in any mining project in the country. Thus, included in the legislation are prescriptions for the government's dual role as a regulator as well as a financial stakeholder for the country's numerous resource extraction projects, including the Frieda River Project.[15]

In mining contexts, the government of Papua New Guinea is enacted by members of the Mineral Resources Authority (MRA). Based in Port Moresby, the Authority acts on behalf of the national government to enforce regulation; provide advice and support to prospective and existing investors operating resource extraction projects in the country; and issue and manage mining licences. For the Project, the MRA was represented by one of its exploration coordinators. I met and spoke with him at the Airstrip Camp during one of his rare visits to the site and visited him in his modern MRA office in Port Moresby. It was during that visit that he entered the room carrying under his arm a number of overfilled, dusty folders and dozens of loose pieces of paper – an archive of the Frieda River Project's long and complex history. It struck me then, and it strikes me now as I write this chapter, that this fragmented and unordered documentation of the Project's development, laid out on the table in the MRA's modern conference room, at once represented and embodied the Project's long and turbulent history. Over the decades that have passed since the deposits at Frieda were first recorded in 1966, these documents had changed hands many times, and had been handled and reorganised by numerous parties, on numerous occasions, and for numerous interests – each of those parties taking from and adding to the account and the history of the Project's development.

The MRA representative was joined in Project-related negotiations by representatives of provincial-level governments, so-called provincial liaison officers – one representing Sandaun (formally the West Sepik Province); the other, the East Sepik Province. Although the two men shared the overall goal of advancing the Project along its mine life cycle towards construction and operation, the professional relationship between them was strained. Frieda's biggest gold and copper deposits, Horse-Ivaal and Nena, subject to an EL-58, as well as all but one recognised impact community were located on the Sandaun provincial territory. As such, in the ongoing negotiations, Sandaun representatives tended to view the mining development at the Frieda River as an exclusively Sandaun Project and hoped to utilise a major share of the Project's benefits within its provincial borders. But the nature of the geographical terrain in which Frieda's deposits are located, in the tributary of the mighty Sepik River, meant that any mining activities would have serious environmental consequences and potentially detrimental effects on the people of East Sepik Province, including Paupe's Paiyamo. The East Sepik provincial representative at the Project firmly stood his ground, demanding that the benefits be split more evenly between the two provinces.[16] He actively advocated that all Sepik communities living downstream from the mouth of the Frieda River ought to be recognised as impact communities of the Frieda River Project and should be consulted in the process of assessing the Project's social, economic and governance feasibility and, should the Project proceed to operations, be included in benefit sharing agreements. In 2011, the company conducted a four-phase awareness campaign that covered fifty-two communities from the junction between the Frieda River and the Sepik River, and the mouth of the Sepik. It reported that the campaign provided 7,800 people living in those communities with 'personal access to accurate information about the project' (Xstrata Copper 2012: 27). While the company acknowledged the need to include people living along the Sepik in what it described as 'stakeholder engagement and information sharing', its community relations and development activities were focused mostly on villages that were formally identified as landowners in the Project area and granted the status of impact communities at the Frieda River Project.

The Impact Communities

In 1995–6 Highlands Pacific conducted a landowner identification exercise in the EL-58 area (see chapter 4). Requested by the Papua New Guinean government, the aim of this exercise was to identify primary landowners of the Frieda River's rich copper and gold deposits (see Jorgensen 2007a and Gardner 2004). The exercise awarded the status of landowner to members of communities located in the Project area: the Telefol villages of Ok Isai and Wabia; the Miyan communities of Wameimin 1 and Wameimin 2, Amaroumin and Hotmin, and the Paiyamo village of Paupe. A Miyan community of Sokamin was later added to this list. These were identified as impact communities, the label reflecting the anticipated and uncertain future impacts of an operating mine as well as the impacts associated with ongoing exploration activities (see chapter 5). Using stakeholder theory, impact communities can also be seen as communities with the power to impact the Project. By the time I reached Frieda, alterations in infrastructural plans proposed by the company meant that Hotmin no longer met the technocratic criteria for the status of an impact community. As a result, the focus of CA activities and community engagement strategies shifted onto the seven remaining villages. Recognition as impact communities awarded them a special status – inclusion in agreement-making processes and a claim to a share of Project benefits. I am not suggesting here that a version of this category was absent from Frieda before the 1996 landowner identification, but it was the labels of 'landowner' and 'impact community' popularised by the 1996 exercise that framed and facilitated the communities' engagement with the Project's other stakeholders that I observed unfolding at Frieda fifteen years later. In the next chapter, I will present the history of Paupe and the process of 'making up the Paiyamo', which largely followed rather than preceded the establishment of the village in 1962. What I want to do here is place Paupe in a broader context of what I came to know as 'the Frieda Community' – comprised of Paupe together with Ok Isai and Wabia – two Telefomin villages located further up the Frieda River, south of Paupe and the Airstrip Camp.

The three communities were distinctive in character. Of the three, Ok Isai was the biggest, with a recorded population of 348 in 2009 (Coffey 2009). Small white planes bearing the logo of the Mission Aviation Fellowship landed regularly on the village airstrip, while the waterfront served as a peculiar and unlikely centre of commerce in this remote and seemingly isolated corner of Papua New Guinea. Most of the buildings in the village were permanent or semi-permanent structures with pained timber walls and copper clad roofs. In the centre of the village, some of the remaining traditional bush material structures, which dominated the settlement's landscape until recently, were used to house petrol-powered generators, which provided electricity to many members of the community. At the waterfront, goods, cash, gold and stories readily changed hands, traders came to exchange cash, petrol and other consumer goods for gold. The relative financial wealth of the Ok Isai community was closely linked to alluvial gold mining, which dominated the local economy.[17] It was an important economic and social activity, but it was largely unregulated. Xstrata representatives told me that they had to be careful not to be seen as endorsing it through the local agreement-making processes or community development activities. It seems somewhat ironic, therefore, that it was one of the mining companies conducting exploration work at Frieda that introduced the area's communities to the process of searching and panning for gold. In the mid-1990s, Highlands Gold Pty Ltd, which later became Highlands Pacific Ltd, employed a number of local people to help gather geological samples, which were then used to assess the extent of the area's mineral wealth and the feasibility of the prospective mining project. Their task was to collect gold nuggets from the river sediment, and so the company taught them how to search for gold in the area's numerous rivers and creeks. Although for the company it was an element of a geological data collection exercise, the community members quickly identified great value in the newly acquired skill. Over time, local extraction methods became increasingly elaborate, with dredging machines becoming more popular, and gold trade became a core part of Ok Isai's economic and social strategy through which the community attempted to assert its own local form of development based on a cash economy and the new kinds of business connections they established with the external traders. A version of this strategy was quickly adopted by the nearby community of Wabia, albeit to a lesser extent. By the time I arrived at Frieda, some members of both communities spent more time in the temporary bush camps erected on the shores of the gold-rich creeks and springs than they did in the villages themselves, and most took part in gold panning trips at least a couple of times a year.[18] Whilst away from the village, these people were commonly referred to as being 'in the bush' – a term traditionally associated with subsistence activities of hunting and gathering but which became popular to describe alluvial gold mining pursuits.

Despite early enthusiasm and promise, with time, the area's mineral wealth and the communities' ability to generate income from alluvial gold mining activities started to be perceived locally as a double-edged sword – a threat as well as an opportunity. As word of a gold rush at Frieda spread around Papua New Guinea, gold seekers from afar made their way to Ok Isai and Wabia. Some of them were gold diggers who wanted to try their luck in the crisp, fast flowing waters of the area's creeks and springs. Then there were the traders, who came to the area to buy gold in exchange for either goods or cash. Many of those traders travelled up the river from the Sepik in motorised dugout canoes carrying dry foods, toiletries and basic household and clothing items. They generally favoured Ok Isai and its busy waterfront over the more distant Wabia, although it was not entirely uncommon for them to make an appearance in this farthest of Frieda's three river communities. In Ok Isai, they rarely entered the village itself, instead choosing to make temporary camps at the waterfront, where they sold their cargo on makeshift stalls. They left the area once all their cargo was exchanged for cash or gold. The traders' presence on the Ok Isai waterfront made it an uneasy space for many people, especially local women and children. Overall, the community came to see the influx of people from other parts of the country as a socially disintegrating force. Under bags of flour and rice, alcohol and marijuana were brought in to the area and consumed at the waterfront in abundance, often causing men to act in a violent, offensive and overly boisterous manner, breaking local rules of social conduct. Although women did come to the waterfront to purchase items or to board one of the many canoes moored in the nearby canal, they did not do so unaccompanied, and the place remained dominated by men. Individuals from even further away and those who preferred more straight forward cash-for-gold transactions would arrive in either of the two Telefomin communities on Mission Aviation Fellowship planes or on motorised canoes to purchase gold to then sell it at an elevated price to buyers in Wewak, Mt Hagen and Vanimo. This particular type of entrepreneurial activity has, over time, created new types of social and wealth stratification in Ok Isai and Wabia, as a few local gold buyers with external contacts purchased gold from workers locally and sold it to bigger vendors, pocketing the difference and making a healthy profit. In many of the village stores in the two Telefomin communities, amongst rice, tinned fish and cans of Coca Cola it was common to see a clearly written note: 'we buy gold for 75 Kina'.

The communities employed a range of strategies to control access to the area's rich natural resources and paid a lot of attention to the movement up and down the river. In evening conversations, people spoke about visitors' arrivals and departures and what they had brought with them and what they took away. The dependency on money and the cash economy in Ok Isai was high, and many visitors to the area (including Paupe's Paiyamo) often remarked on the fact that relatively few local community members cultivated gardens and even fewer continued Telefomin's rich and elaborate hunting tradition. Although these processes were less severe in Wabia, members of both communities shared with me their worries about the impact that the influx of cash and people into the area has had on the integrity of local communities and *kastom*. They were also concerned that the informal gold mining sector had introduced a new type of politics of inclusion and exclusion within the two communities and felt that not everyone had equal access to the benefits from this peculiar and dynamic form of local economy. These experiences informed their (as well as the Paiyamo's) thinking about the Frieda River Project and the possible negative consequences of large-scale resource development. Despite growing concerns about immigration, however, the support for the Project in all three communities remained high.

In many ways, Paupe differed from its two Telefomin neighbours. It was the smallest of the three impact communities located along the banks of the Frieda River, with a population recorded at 163 during a local survey in 2009. It bore relatively few visible signs of the version of development pursued by its southerly neighbours. When I first arrived there in 2011, there were only a handful semi-permanent buildings in the village: a Baptist church with concrete floor and a tin roof; an old, unkempt company outpost; an old house built for company visitors, which Highlands Pacific Ltd presented to pastor Jacob; and the remains of a company clinic, which marked the entry to Paupe when approached on foot or by boat from the airstrip. All remaining buildings were temporary structures made of bush materials and roofed with meticulously woven sago palm leafs, which provided welcome shade and refreshing, cool air inside on a hot, sunny day. The Paiyamo were by no means isolated and negligent of the new type of economic activities and relations that had gained prominence further up the river, and most members of the community would sporadically call on their widely defined kinship connections to access these opportunities in Ok Isai and Wabia. Although I often heard them talk about the abundance of gold in the creeks surrounding Paupe, the Paiyamo seemingly felt no need to re-create more locally the opportunistic model of economic development employed in the two Telefomin communities. Instead, they took pride in the fact that they maintained a largely subsistent lifestyle and held on to what they called their *kastom* whilst enjoying access to goods, transport and services through their engagements with the company personnel and occasional visits to gold panning hotspots near Wabia or Ok Isai. Few of the canoes that passed the village on their way up to Ok Isai and Wabia ever approached its sandy banks, but the Paiyamo made frequent and much enjoyed visits to the Ok Isai waterfront, combining shopping for essential (and less essential) dry goods with visiting their friends and kin. Like their neighbours in Ok Isai and Wabia, they actively embraced the label of 'impact community' introduced in the context of the Frieda River Project, and descriptions of a bright if unknown future featured prominently in evening conversations by the fire.

Categorical Dissonance

It is no coincidence that the emergence of the current version of the impact community label at Frieda, together with its ascribed roles and characteristics, was linked closely to the 1996 landowner identification – an administrative exercise aimed at finding and recording distinctive units of land and their owners. After all, for all intents and purposes, the land and the minerals buried deep underneath it was precisely what brought the communities, the company and the government together at Frieda as stakeholders to the Frieda River Project. More specifically, what brought the stakeholders together at Frieda was the transformative potential they saw in the copper and gold deposits. Their aspirations were frequently expressed in terms of development. For the company, constructing and operating a mine at Frieda would mean delivering a successful resource extraction project, growing Xstrata's portfolio, expanding its market share and developing as a business. For the national government, the anticipated mine at Frieda River would entail a boost to its Gross Domestic Product and a series of other economic and human development indicators. For the provincial governments, the mine at Frieda bore a promise of delivering much needed infrastructure and services to areas where they were otherwise lacking, creating training and employment opportunities and contributing to the provincial budgets. For

Illustration 1.4. Riverbank in Aston. Photograph by the author.

the impact communities, the promise of bringing (mining) development to Frieda bore within it a vision of different, more prosperous lives that could be lived along the Frieda River.

This context ascribes a number of meanings to the phrase 'developing Frieda'. The phrase can be used here to describe a process through which the Frieda River Project progressed through the Mineral Project life cycle with the hope of eventually becoming an operating mine. But it can also be used to capture the process of bringing more broadly conceived development to the Frieda area. Depending on the angle and the type of conversation, the mine at Frieda can therefore be seen as a result of, the trigger for, and the embodiment of development itself. If development at Frieda can be seen as an aim and a successful outcome of stakeholder relations, it can be argued that stakeholders' ability to bring this development about was linked to their ability to perform their ascribed roles and form effective stakeholder relations. In line with ascribed roles, the company was both the investor and technical operator of the Frieda River Project. It held the capital resources, expertise and technological capacity to extract gold and copper from Frieda's deposits. It operated within the context of legal frameworks and agreements set out by the government – the project's regulator (although, as noted above, the government also held significant financial interests in the Frieda River Project). The company's engagement with people living at Frieda defined them under the label of impact communities, a seeming demonstration of acknowledgement of the Project's externalities and its environmental and social impact, but in reality a fulfilment of its legal obligations towards the state and the Project's formal landowners, linked to an attempt to limit its engagement and liabilities to a definable population. To assert their role and maintain a relationship with the Project and its stakeholders, the Paiyamo, alongside other impact communities, had to successfully claim ownership of the land at Frieda and use their intellectual and customary repertoire to produce evidence in support of this claim. On-site, in the mining locality and at the Project's negotiating tables, all the above roles were enacted by individuals and brought to life through the medium of social relations.

In the introduction to her book *Kinship, Law and the Unexpected: Relatives Are Always a Surprise*, Marilyn Strathern (2005: 7) identified two kinds of relations which, she suggested, come together to create a tool for social living. The first kind, she claimed, are conceptual or categorical relations, connections acquired 'through a logic or power of articulation that acquires its own conceptual momentum'. The second kind are interpersonal relations taken as connections that develop over time, 'inflected in a precise and particular history'. Strathern brought the two together and argued that 'it is through interacting with persons that diverse interactions and further connections become intellectually conceivable, while it is through creating concepts and categories that connections come to have a social life of their own' (ibid.: 8). Using the interplay between categorical relations prescribed in stakeholder roles and interpersonal relations formed by people enacting those roles on the ground I aim to show here how the two came together and awarded a social life to the Frieda River Project. What emerges is what I term here as 'categorical dissonance', which contrasts expectations of perceived roles and responsibilities with actual enactments and effects of those roles. The concept pivots on a cognitive dissonance theory first proposed by Festinger in the late 1950s, which 'suggests that when individuals hold two or more cognitions that are contradictory, they will feel an unpleasant state – dissonance – until they are able to resolve this state by altering their cognitions' (Hinojosa et al. 2016: 171). The ethnographic material presented here speaks to two interconnected types of categorical dissonance.

The first is dissonance within categories, found in dynamic spaces between categorical prescriptions of stakeholder roles and responsibilities and their enactment in project contexts. It will be explored in chapters 2–4 by looking at ways in which the Paiyamo strived to understand and perform, to the best of their ability, their role as a landowner and impact community of the Frieda River Project. This identity changed constantly as communities played off each other and responded to what they perceived were the roles expected of them by the company and the government. The book contains several stories that the Paiyamo told to explain the Project and to assert their role within it. In one way or another all these stories present the process of planning for mining development at Frieda as a relational quest. They offer explanations for different roles and relations that the Frieda River Project has mobilised over time, and they put to the test a series of assumptions about stakeholder categories and their enactments at resource extraction projects. A version of intra-categorical dissonance was present among all key stakeholder groups at Frieda. More often than not, people representing the company and the government at the Project were acutely aware that they were partaking in engagements riddled with ambiguity and incongruities and that required groups and individuals to perform roles that, contrary to appearances, did not have an independent existence. To understand those processes, it is necessary to pay attention to the broader picture and to the blurred lines between what the prescribed roles say and what is actually happening when different categorical relations and cultural logics meet and become animated. For example, most company personnel interviewed for this research talked about feeling trapped between company procedures and performance indicators and the realities of their daily work (see chapter 5). To perform their roles, they formed relationships with members of local communities as well as the national and provincial government representatives embedded in Xstrata's strategies and procedures. In practice, however, the terms of these relations were being constantly negotiated between individuals actively involved in the process of planning for mining development at Frieda.

The second type is dissonance between categories, which emerges where one or more stakeholder groups fail to meet the expectations of their role, leading to absences in their respective functions. Where such absence arises in a way that threatens the shared project, another actor (or actors), with the capacity to do so, will likely fulfil the missing role to address the disharmony and ensure continuity of the process – in this case, moving the Frieda River Project towards construction and operations. This will be illustrated here by looking at the state's performance against prescriptions of its categorical role in the Frieda River Project. To that end, chapter 5 describes how the company developing the project has attempted to mediate the failures of the government, or its profound absence, and found itself under pressure to take on service delivery functions in the Project area. A reflection of inter-categorical dissonance can also be found in ways in which the company and the government 'stepped in' and manufactured local categories to serve their own means of Project governance and to allow communities to become 'legible', to borrow a term from Scott (1998), and grant them the status of stakeholders of the Frieda River Project. Stakeholder categories of the type described here do not just emerge but are strategically created to serve a given purpose.

Papua New Guinea's track record in mining highlights the great economic as well as political power accumulated by resource extraction companies (Kirsch 2002; Imbun 2007), including the power to create and utilise labels and categories. But, although 'the balance of power between companies and local communities might be uneven . . . the relationship has never been unidirectional' (Bainton 2010: 14). At Frieda, local communities were dependent on the state to create a regulatory environment that would simultaneously facilitate and hold to account the company's activities and generate a legal space within which they themselves could claim official recognition as landowners in the Project area and, therefore, assert a position among the Project's stakeholders. The government and the communities both relied on the company's technological and economic capacities to turn the gold and copper minerals into broadly conceived development, while the company relied on the government's ability to implement an effective governance framework for the Project and the communities' permission to access their lands for exploration and, eventually, extraction activities. Each of those actors used their own ontology to frame, assert and enact their role in the process of planning for mining development at Frieda and, as will be shown here, the readings they made of each other informed their expectations and subsequent actions.

The systemic power inequalities mean that, in this process, the different stakeholders were not equal partners – and it would be naive to think otherwise. Taking stock of power inequalities in the global mining industry, Owen and Kemp (2017: 8) described a process through which mining companies applied an extractive approach to social as well as physical environments hosting mining projects. This included 'the way mining companies engage and recognise landowner groups and their interest in the development process, configure their social investments over the mine life cycle, interpret the interests that drive and later sustain negotiations and agreements with local communities, and management of impacts and legacy issues'. However, as Owen and Kemp point out, in a reflection of this process, mining also 'creates opportunities for different individuals and organisations to extract a share or a benefit from the mining business'. What I suggest here is that although extraction of the type described by Owen and Kemp is inevitably an inherent part of the mining industry, for the Frieda River Project, stakeholders came together not only to extract but also to create and utilise resources, driven by their shared interest of precipitating mining development at Frieda.

I have no doubt that the representatives of all three interest groups included in this study were aware of the power inequalities that underpinned stakeholder engagement activities at the Frieda River Project. In direct conversations, however, they used a range of strategies to claim ownership and control over the future of the Project. In one of our conversations, a provincial liaison officer told me that the Frieda River Project belongs to the government. He strongly implied that although the impact communities must be consulted, ultimately, the destiny of the mine at Frieda lies in the hands of the government and the company. Together, he said, they have the power and the capacity to proceed with the plans for the mine at Frieda even against communities' wishes. He was adamant that the company ought to engage with the government more closely, as this was the only way to ensure longevity of the venture and sustainability of development at Frieda. 'Government will stay here for life. The company will leave,' he told me. But this statement, although in principle true, crumbles under closer investigation. Although it is reasonable to assume that a version of the government will 'stay at Frieda for life', the people who enact the policies and priorities of the government and their relationships with the Project's other stakeholders are more than likely to change over time. In 2011, the upcoming National Elections and uncertainty those elections entailed regarding the Project's regulatory and operating environment were a prominent feature of conversations in company offices at the Airstrip Camp. The Paiyamo were also weary of the government, viewing it as a powerful but uncertain and ambiguous creature with whom they shared only passing relations. Instead, they used their newly adopted land ownership ideology to assert for themselves a core position in the nexus of relations at Frieda, playing the government at its own game of consequential and powerful labels and categories, as the material presented in the following chapters will show.

In the process of planning for the Frieda River Mine, the impact communities were represented by community spokesmen. The spokesmen were chosen by the people but appointed by the company. Although the form of community representation was facilitated by Xstrata, the CA officers had serious concerns regarding the quality of representation at Frieda and the extent to which the appointed representatives actually represented the interests of their communities. Especially in the two Telefol villages, socio-economic processes associated with alluvial gold mining led to the emergence of a number of powerful individuals with strong, external connections who, the officers suspected, brought vested interests into agreement making. They feared that, in the long term, this stood to have a negative impact on the quality of relations with the communities affected by the Project. In principle, however, around and away from the negotiating tables, members of the local communities could often be heard saying that of all the agents involved in the agreement making, the communities themselves and their leaders were the ones who, when the time came, would ultimately make the key decisions regarding the future of the Frieda River Project and facilitate the big reveal of a mine at Frieda.

Xstrata entered this set of relations and speculations from quite the opposite angle. Far from claiming the status of a constant element of the Project, the company consciously framed its engagement and relations with the other stakeholders around the notion of fixed and limited temporality. At the time, a full feasibility study of Frieda's deposits was nearing completion. As I describe in chapter 5, Xstrata's stakeholder engagements were strongly influenced by the fact that no decisions have yet been made regarding their future involvement with the Project. As such, the company has carefully managed its relationships with the government of Papua New Guinea and the impact communities in such a way as to be able to exit those relations at relatively short notice and with minimum liabilities should the Project not progress to construction and operation under its management. From the perspective of Xstrata, it was they who maintained control of the situation. As described above, they were a catalyst for the relations framed around Frieda's deposits. The main reason behind this sense of control and relative security at this stage, however, stemmed precisely from the company's largely non-committal involvement in the Project. Maintaining a version of this non-commitment, I argue, meant that the company was clear about differentiating between the current Project and the potential future mine. To fully understand not just the appeal of but also the possible need for such a distinction, it is important to consider that although administratively Xstrata was a fixed business unit that could be influenced and managed through processes prescribed by its legal, financial and bureaucratic set-up, the company shared a lot of fluid characteristics with the community and the government. The roles and responsibilities ascribed to the company personnel were in a constant state of flux, with business priorities and activities influenced by changing organisational models as well as fluctuating stock markets and commodity prices.

Conclusions

Anthropology has long played a part in revealing the ambivalence and contradictions of the 'landowner' label, approaching stakeholder relations in terms of a complex and dynamic network of mutually constitutive roles and expectations animated by a wide and varied matrix of social processes and interpersonal relations. More recently, the discipline has applied a similar level of academic scrutiny to 'the government' and 'the state' as stakeholder categories at resource extraction projects. The results paint a picture of stakeholder relations as riddled with ambiguity of form, carefully designed to simplify and create a version of reality that seems prescribed and factual when they only really exist, if at all, in very particular contexts, negotiated through complex categorical and interpersonal networks. At Frieda, the anticipation of a large-scale copper and gold mine is what brought the different stakeholder categories to life. At the same time, the existence of a mine at Frieda became dependent on the different stakeholder groups' ability and willingness to enact their roles in a way that would pave the road for the mining development to take place. In other words, in the recognised presence of mineral deposits in the area, it was down to people and groups involved in the process of negotiating and enacting stakeholder relations at the Frieda River Project to make the mine at Frieda happen, making the mine an inherently relational undertaking.

The network of stakeholders at the Frieda River Project was and remains broad, encompassing mineral consumers and market regulators; corporate shareholders and executive officials; in-country planners and industry bodies; nongovernmental organisations and watchdogs. But the events presented here, although embedded in those broader spheres of power, interdependencies and influence, unfolded in the location of the proposed mine, on the banks of the Frieda River, where stakeholder roles were enacted and brought to life by the medium of social relations.

Notes

1. See also Blair (1998, 2010) and Reed et al. (2009).
2. This definition has since been criticised for reducing stakeholder roles to purely prudential interests (Phillips 1997; Phillips, Freeman and Wicks 2010) and was expanded to include a concern for the wellbeing of the parties affected by corporate decisions (Cragg and Greenbaum 2002). Crucially, it has been linked to the recognition that a company is a 'web of relationships among the stakeholders' bound together by reciprocal regimes of responsibility (Goodstein and Wicks 2007: 395; see also Donaldson and Preston 1995).
3. The stakeholder theory has been rebranded as 'the names and faces approach', shifting the research focus away from largely dominant strategic decision-making towards individual relationships. This shift in emphasis took place amongst acknowledgement that the stakeholders themselves 'have begun to challenge the borders between traditional stakeholder groups', making it increasingly difficult to put them into neat categories (McVea and Freeman 2005: 62) and even more difficult to consider them as objective entities existing independently of the complex network of social, legal and economic relations that animate them.
4. See, for example, Jorgensen (1997); Hirsch (2001c); Leach (2004); see also chapters 2 and 4 here.
5. See also Hirsch (2004) for his account of the process of entification among the Fuyuge- speaking communities in Papua New Guinea's Central Province.
6. See, for example, Gilberthorpe (2007) and Weiner (1995, 1998b, 2001, 2002, 2007) for examples of landowner identification for the Kutubu Oil Project.
7. Scott's (1998) model describes the process in which state administrations invent and implement specific categories in an attempt to impose order on society and nature.
8. There was no military presence in the Frieda River area during the Second World War, but this statement suggests that stories of army combat reached Paupe before the arrival of the CEC.
9. In the 1960s and 70s, John Pasquarelli owned a trading company in Papua New Guinea with trading posts in Ambunti, Amboin and May River. In 1964, he became a member of the Papua New Guinean parliament. He was also a collector for the Basel Museum, and it was on a shield- finding mission in 1963, in his role as a collector, that, he claims, he reached Frieda and was the first to discover traces of gold and copper mineralisation. The extract above comes from his personal notes, which he kindly shared with me in 2014. A version of these notes has previously been published in *The National* in March 2012 in an article entitled 'How Frieda River copper and gold was discovered'.
10. What I saw was not the original document but a copy requested by Tony Friend during a 1996 land determination exercise, which he presented to the Paupe community.

11. The schedule includes the names of twenty-five owners, both male and female, twenty-two of whom were recorded as residents of Paupe village (with different places of residence given for the remaining three).
12. Tikap was an Iwam speaker married to a Paiyamo woman. He was considered a Melefale elder and a prominent figure in Paupe, where he settled in the late 1960s. He told me that at the time of the sale his knowledge of Paiyamo Amoke was limited. He did not fully understand what the white man was saying and, although he did his best, he was not able to explain things clearly to Aiyapanai.
13. Further insights into this complex process can be found in Weiner's argument that 'traditional custom cannot be protected by an act of legislation. The legislation is composed and empowered by a cultural and legal system very much at odds with the way local "traditional custom" arises and is implemented' (2001: 21; see also Filer 2011, 2012).
14. As shown in Table 1.1., the Frieda River Project was co-owned by Xstrata Frieda River Ltd and Highlands Pacific Ltd between 2007 and 2013.
15. For a history of mineral development policies in Papua New Guinea, see Filer and Imbun (2009).
16. In 2012, I worked as a Domestic Observer at the country's 2012 Parliamentary Elections. Assigned to one of the Ambunti District's electoral teams, I visited a number of Sepik communities located downstream from the mouth of the Frieda River. Everywhere I went people would ask me about the mine. They would tell me that they did not want the mine to operate, as the damage it would cause to the Sepik River and, inherently, to their livelihoods would be irreversible. At the same time, many adult men and some women from those communities would discreetly probe me about access to services and employment opportunities at Frieda. The general feeling was that the East Sepik Province and the Sepik communities should be included in the process of planning for mining development at Frieda. Through a variety of means and tactics, people in those Sepik villages tried, unsuccessfully, to claim a version of a relationship they heard of and observed between the company and the officially recognised impact communities of the Frieda River Project, hoping that this would grant them access to opportunities they felt were opening up for their neighbours along the Frieda River.
17. Although alluvial gold mining in PNG remains a subject largely unexplored by anthropologists, Moretti (2006) offers an interesting account of artisan or small-scale mining in Mount Kaindi, Papua New Guinea.
18. The demand for income was particularly high at Christmas time and the start of a new school term, when families needed to find sufficient money to cover the school fees.

Illustration 2.1. A large dugout canoe. Fishing trip to one of Frieda's numerous creeks. Photograph by the author.

2

Making up the Paiyamo

Introduction

There was a house in Paupe that was not like any other. Its floors were dusty and the sago leaves on its roof were beginning to crumble. But there was an undeniable sense of grandeur about it. It stood high, raised above the ground and perched on top of a hill in the southernmost part of the village. The Paiyamo referred to it as the big house or the *kastom* house. When they talked about it they did so with respect and, at times, even with a touch of melancholy. But there was also a strange sense of detachment in the way they related to the building. The house, I was told, represented the kind of lives they lived before the company arrived at Frieda, before a time of *kastom* and ancestors gave way to what they referred to as 'the time of white people'. Back then, they explained to me, they were not known as the Paiyamo Sane[1] but were several different groups. They shared one language, Paiyamo Amoke, but lived separately in longhouses scattered around a large area along the banks of the Frieda River. This chapter tells the story of how those groups came together and became a Paiyamo Sane community of Paupe and landowners in the Frieda River Project area.

The title of this chapter is borrowed from Ian Hacking's work on 'Making up People'. In his book, Hacking (1986: 161) described a process through which the nineteenth-century preoccupation with proliferation of labels in the language of social control (both academic and political) enabled the process of making up what he referred to as new 'kinds of people'. This process, he claimed, was made possible by social changes that carried with them new forms of description and therefore 'new ways for people to be' as these descriptions brought into being novel possibilities for action. This chapter describes how the colonial administration created a village on the banks of the Frieda River and categorised the Paiyamo speakers it brought together as 'the people of Paupe', and how the subsequent land identification exercises introduced the labels of 'landowner' and 'landowning clan' into the local repertoire, indeed creating new ways for people to be in a complex system of influence and control, and in the contested context of the Frieda River Project. The Paiyamo were not passively altered into landowners but actively transformed themselves in the hope of affecting a range of current and future opportunities. Local temporalities play an important part in those processes. The chapter begins to unravel the Paiyamo understanding of time, in which present and future complete the past and explanations of the past can be altered to reflect current aspirations and serve future-oriented pursuits. In Paupe, where no event can be fully knowable at the time of happening, its meaning and significance only ever revealed over time, the resource exploration project opened a discursive space that revealed new meanings and power in the events and stories of old and created novel ways of effecting and affecting change. The Paiyamo skilfully utilised this space and filled it with a repertoire of narratives. While the official documentation places the origins of the Frieda River Project in the late 1960s, this chapter presents the local history of the Project, as told by the Paiyamo, reaching all the way back to the time of ancestors, land spirits and *kastom*.

Stories in Time

I got to know the history of the Paiyamo and their history of the Frieda River Project not through a continuous narrative but through a number of idiosyncratic stories. Each of those related to a period in time resembling what Wagner (1986) would call an epoch – a distinctive manifestation of time's presence rather than an indicator of its passage.[2] Each epoch described to me by the Paiyamo was unique and began with what can be described as a rupture in their world when a new arrival in the area introduced new qualities into the local network of relations, creating novel opportunities for the types of lives that could be lived on the banks of the Frieda River. The stories spoke about the time of the ancestors and *kastom*, the time of Christianity (see chapter 3) and about how the arrival of colonial administration and the first mineral exploration company marked the beginning of the time of white people. This manner of speaking about history and time, as a series of distinctive epochs, is relatively common in Melanesia. While a range of 'western' ideas about progress, advancement and development seem to adopt an evolutionary view of time, where social change occurs through a gradual and continuous process, numerous anthropological accounts from the region suggest that there exists, in Papua New Guinea, a very different, episodic (or neo-episodic) way of moving through time that entails distinctive continuous periods separated by moments of radical and complete transformation.[3]

Much has been written about the episodic view of time in the Sepik contexts. Simon Harrison (1982: 147) used an annual cycle of yam gardening to explain the symbolic representation of time among the Manambu. Borrowing a phrase from Evans-Pritchard, he argued that 'people tend to think of time as a relation between activities' and showed that 'at Avatip these activities [were] essentially those connected with yam horticulture'. Also focusing on a cycle of ceremonial yam growing, Richard Scaglion (1999: 212) explained how 'the Abelam take an episodic view of history, in which the flow of time produces "punctuated equilibria"; steady states ruptured by catastrophic events that restructure the world and result in new steady states'. He argued that 'assuming a convenient fiction of a traditional Abelam person he or she lives in a sort of timeless present in which no change is possible' (ibid.: 218) for real change 'can only occur cataclysmically via a total restructuring of the world' (ibid.: 211). In a similar realm, Nancy McDowell (1988: 124) described how in Bun people 'do not conceive of the past as a series of interconnected events in a cause-effect chain. For them there is no gradual, cumulative, evolutionary change; change is always dramatic, total and complete'. Like Harrison and Scaglion, she too argued that in the neo-episodic view of time 'change occurs only through discontinuity'. Listening to the Paiyamo tell their stories it is really not difficult to see the appeal of the dualistic distinction between the evolutionary and the episodic experiences of time first proposed by Gellner in 1964. The Paiyamo make the different epochs seem easy to identify and clearly articulate the ruptures marking the points in time at which one story and one state of being ends and another one begins. But upon closer inspection, when the prescriptions of an episodic model are applied to the everyday lives of the people, cracks inevitably begin to appear. Framed in relational terms, where this frame encompasses a range of social as well as structural relations, the way in which people relate to land and their surroundings, and the interaction between the visible and the invisible realm, such a view of temporality only tells part of the story.

Borut Telban's work offers valuable insights here. In his description of what he called an egalitarian temporality among the Karawari-speaking Ambonwari of the East Sepik Province, Telban contested the principle of complete rupture and discontinuity as a prerequisite for transformation from one state to another. 'While Melanesians generally anticipate immediate and complete change – be it the result of initiation ritual, religious movement, or economic activity – the actual change rarely meets people's expectations' he wrote (2017: 195). For Telban, the neo-episodic model of temporality failed to fully capture the experiences of the Ambonwari for two main reasons. Firstly, the model was based on a falsely homogenous view of the ways that people and communities move through time and experience change, as any collective temporality contains multiple temporalities of different groups and people and is 'filled with heterogeneity, competitiveness and changeability' (ibid.). Using his ethnographic data, he portrayed the Ambonwari as a community where different beings, belonging to both the visible and invisible realm of the world, were engaged 'in a continuous struggle for prestige and influence over others and their attempts to reach a temporary consensus that would create a unifying period' (ibid.: 187). This puts at least some of the responsibility for creating periods and affecting change on the Ambonwari themselves and organises agentive time around future-oriented projects. Secondly, building on the works by Bateson (1958), Schuster (1990) and Wassmann (1990), Telban argued that people's history is inherently linked to local geography, tying time and being in time to a place, thus making the latter 'a mnemonic device for past events and relationships' (ibid.: 193). Together, he claimed, the emplacement of people's temporalities in the surrounding landscape and the coexistence of multiple temporalities in the community formed 'constant barriers to people's aspiration for a total transformation' (ibid.: 194) of the kind described by Gellner.

At Frieda, the prospect of mining development and the emergence of stakeholder categories mean that a new rupture has occurred and a different kind of transformation has taken place. In the time I spent with them, the Paiyamo continued to talk about their history in terms of distinctive periods of time separated by ruptures in the fabric of the world as they knew it, but they would often say to me that they were only just now able to understand the true meaning and efficacy of the old stories and past events they described. Like when they concluded the story of Afek and *bangri* by saying that 'it is an old story but now things are happening and the story is becoming true. It is an old story but it is true now' (see introduction). Rather than requiring total discontinuity and renunciation of everything from the past, this most recent transformation created a space in which Frieda's communities were able to draw power and capacity in the present from the stories seemingly belonging to the previous epochs and the time of *kastom*. In this process, emplaced myths and multiple temporalities served not as barriers to Paiyamo's aspiration for transformation, to borrow Telban's turn of phrase, but as vehicles for affecting and effecting desired change.

One element of Gellner's theory seems particularly relevant to the Paiyamo and can aid our understanding of the way they experience their past and imagine their future – the way the model frames people's anticipation of change. As described by Nancy McDowell (1988: 124–25), 'the episodic conception of history and change is not restricted to the past but provides a model for change in the future as well: coming change must also be total, drastic and radical'. She argued that 'because people perceive change to be of this nature they have deep and profound expectations about how change will come about. They expect dramatic revolutions; if one thing changes, everything will change – they only need to find the key(s) for controlling and directing the change, preventing or affecting it'. McDowell argued that this kind of expectation itself was not new for the people she worked with. What was new were 'all the visions about possibilities for the future' (ibid.: 125). The way in which the Paiyamo tell their stories of the Frieda River Project, some of which are presented in this book, offers glimpses into the ways they imagine what this future might be for them and the way they conceptualise how 'change can be effected as well as affected' (ibid.: 122). They do fully expect the kind of dramatic change described by McDowell and can often be heard saying that when the mine at Frieda River reveals itself and when development comes their lives will be completely transformed. To that end, it seems to me that one of the key messages conveyed in the stories Paiyamo tell of the ways they were, the ways they are and what events marked a transition from one state to another is one of their sheer capacity and readiness for transformation and, in the context of this research, for mining development at Frieda River.

New Kinds of People?

The previous chapter described how the Project introduced stakeholder categories at Frieda. It problematised the notion of stakeholders and presented them as products of complex and dynamic relations, realised in the interplay between the categorical prescription of stakeholder roles and their enactment through interpersonal relations negotiated between people representing key interest groups at the Frieda River Project. Studied through the lens of Gellner's (1964) theory of neo-episodic time, such externally introduced forms of description can be seen as mechanisms for precisely the kinds of transformations through which people can move from one state of being to another and become 'of a new kind'. In this scenario, change is enabled by an introduction of outside forces and concepts into local cosmology, which is often seen in anthropological accounts from Melanesia as a prerequisite for transition from one episodic state of being to another (see, e.g., McDowell 1988; Scaglion 1999, 2000). Errington (1974: 257) argued that such external influences are necessary for change because, in the neo-episodic view of time, 'history is not seen as a process of development in which the past contains the seeds of the present and the future'. But this is not necessarily the case among the Paiyamo, for whom much of the future happened in the past but has not yet fully revealed itself and where stories of the past hold the key to unlocking the multiple aspirations for the future. There, on the green banks of a fast-flowing river, forces from outside the area were posing new opportunities, and change had been occurring long before the German woman arrived from the Sepik. For example, as I describe below, the local cult was contingent on inducing and enabling vivid transformations (e.g. from being a male child to a member of the cult) and invoked a great potential for change and innovation, which the Paiyamo skilfully utilised. When the external influences materialised in the form of colonial and then state administration, missionaries and the mining companies, what fascinated the Paiyamo was not that these encounters produced opportunities for change (for it was something they were familiar with). It was the new type and the sheer unprecedented scale of those transformative openings. In a relatively short period of time, through the economic, political and social relations that make up the Frieda River Project, the Paiyamo found themselves caught up in a complex network of interdependencies and relations that were not bound by boundaries of territory, language or race. Over time, as landowners in the Project area, they adopted new forms of description and became, in Hacking's theoretical view at last, 'of a new kind'.

One response to Hacking's theory is particularly useful in the context of this study. It comes from Eric Hirsch, who used Hacking's model of making up kinds to argue that in the face of colonial, missionary and capitalist introductions people in Papua were entangled in a process of becoming people of a specific kind and devising new ways to be a person. Following Hacking, he claimed that 'a new world of possibilities becomes available once new concepts, new descriptions are present' (2001a: 244) and linked the process of generating human 'kinds' to the process of colonisation and a 'manner of classifying that has its origins in (industrial and state) bureaucracies' (ibid.). Several comments made by Hirsch merit addressing here. Hirsch (2001a: 241) argued that in the colonial process of making up kinds 'Melanesians were not simply passively altered, but actively *transformed themselves*' (original emphasis), incorporating and embracing the potential encoded in new forms of descriptions into local discourses and coercions.[4] These processes tended to be localised and informed heavily by local contexts as well as previous experiences of transformation. As Stewart and Strathern (2002: 368) point out, 'we need not suppose that the idea of "kinds of people" was an entirely new colonial imposition, although colonial circumstances often produced sets of these kinds that were different from those which have been held before'. Using examples from their work among the Duna of the Southern Highlands Province, they argued that the seemingly new transformations 'may often have reflected older versions of similar transformations belonging to the past' (ibid.: 367).[5] The Paiyamo stories recounted below and in the following chapters describe a number of such transformative processes. While the narratives are diverse, they all follow a similar pattern, taking their beginning from the moment when an unknown agent (human or supernatural) entered their area and, in one way or another, presented them with new opportunities and novel ways of being in the world (as distinctive lineages, as the people of Paupe, as Papua New Guineans, as Christians, or as landowners in the Frieda River Project area – just to name a few).

Further, speaking of new forms of description and classification, Hirsch (2001a: 244) proposed that 'the operation of bureaucracies is guided by the convention that interventions into the lives of people can only be made if it is known what *kind* of person is being dealt with' (original emphasis). Applied to the Frieda River Project and resource extraction projects across Papua New Guinea more generally, this would suggest that the introduction of new forms of description and categories (such as 'landowners') into a complex network of social relations was part of state and successive companies' effort to get to *know* local communities and to formulate particular kinds of relations with them, although note that a very particular kind of intersubjective knowledge is at play here. Described in those terms, the process of 'making up kinds' brings to the fore the asymmetrical power relations between what Hacking (1995a) termed 'the known' (that of a kind) and 'the knower' (that who identifies the kind in the first place). What I want to show in these pages is that, despite structural power inequalities that prevail, through the medium of social relations this process takes a different form at Frieda – one in which the known is also a knower. The relational context is crucial here as it is through experience and understanding formed on the basis of a social relation that the knower becomes the known himself and the known is actually 'a community of knowers' (Wagner 1981: 13). This is not meant to underplay the differences in the ways in which the different groups and individuals at the Frieda River Project understood and practised knowledge and what they took to be knowledge capable of causing effect in the world (see Leach 2012a and 2012b). These differences were stark in the process of planning for an operating mine at Frieda River – as will be discussed in detail in chapter 4.

Building on Gellner's theories of neo-episodic time, Hirsch (2001a: 245) also argued that 'when persons are compelled to change from one state to another, recognition and accomplishment of the transformation relies on manifesting "skin" in the appropriate manner', where 'skin' entails a visible demonstration of a social phenomenon. To achieve such transformation, entities 'must be aligned in such a manner that a unified appearance can be achieved' and they need to have 'the capacity to reveal what originates elsewhere from the context of particular relations' (ibid.: 246). The unified appearance, Hirsch argued, requires multiple interests to align to a single purpose, with the successful revelation of a collective appearance of a 'kind' dependent on relations through which the power and efficacy of it can be demonstrated. At Frieda, the prospect of a large-scale mining development delivered both. It offered an incentive for new types of collective appearance (e.g. as landowners), in some situations overriding diverse individual interests and aspirations present within communities. It also created a social, economic and political context within which these new forms of collective appearance could be made effective. But it was not a straightforward, linear process, as the forms of description induced by the company, and to a lesser extent the government, in the context of the Project became entangled with a whole array of local dynamics and processes, resulting in an end product that was in a constant state of flux and that resulted in neither the elicitation of facts about the Paiyamo as a people or a group nor the total transformation of their sociality as imagined by Gellner.

There is one final aspect of Hirsch's work on making up people in Papua that needs addressing here. Responding to comments about his original piece, Hirsch (2002: 369) argued that 'the way in which a people conceptualize "the past" sets limits on what can be envisioned as their future possibilities. At the same time, and as an effect, this has a direct bearing on what is perceived as their appropriate present courses of action'. He claimed that for the Fuyuge, conceptualisation of the past both informed and imposed limits on their present and future patterns of action. While I agree that conceptualisations of 'the past' do inform people's current actions and future aspirations, in this statement Hirsch seems to have overlooked a critical element of this process – one in which the conceptualisations of the past evolve and can be changed to validate and direct people's current actions and expand rather than limit their opportunities for the future.[6] At Frieda, storytelling offered an important vehicle through which the past could be reframed as to offer an alternative explanation of the present and create new possible futures.[7]

The (Hi)Story of the Frieda River Project

At Frieda, uncovering the (hi)story of the Frieda River Project was a highly palpable process that drew attention to 'the political contest between competing versions of the past' (C. Stewart 2016: 89). The stories told of the origins of mineral deposits at Frieda, the Paiyamo ancestors and *kastom* gained new efficacy in the context of asserting positions in relation to the Project. These stories moved freely through time and changed shape as new details and meanings came to light, and as the past was reimagined and transformed into a new version of local history. In his speech at the Pacific History Association conference in Hilo in July 1996, Greg Denning (1997: 420) argued that this is as much as historians (and, by extension, those interested in the study of history) can ever hope to achieve, for they can never actually observe the past but 'the past transformed in some way into history'. In his earlier work, he wrote that 'the past is never contemporary, but history always is. History is always bound to the present in some way. History always represents the present in the ways it re-presents the past' (1989: 134). This is why the Paiyamo stories presented in this book should not be read simply as a factual account of their past but as a contemporary version of their history, reflecting and affecting a whole range of past, contemporary and future circumstances and relations.

Writing about Oceanic historicities,[8] Chris Ballard reaffirmed the long-standing argument that local history cannot be separated from its contemporary context and representation, and the stories I heard at Frieda and that came together to tell the history of the Paiyamo and the Frieda River Project are no exception. History is always implicated in the politics of its time and is never just history – it is always 'history-for' (Levi-Strauss 1966: 259 quoted in Ballard 2014: 108, see also Sahlins 1983 and 1985). To make the most out of the ethnographic material presented here, it must be read within the context of preparing for and precipitating mining development at Frieda. Ballard (2014: 96) argued that histories make the past present but are themselves fundamentally cultural artefacts in that they are 'profoundly performative – they exist in the present and are remade in the act of communication'.[9] Crucially, he remarked that '[i]f historicities direct our attention to the present, to questions of who is telling the history, and why, and for whom, they are also grounded firmly in a sense of the future. Different temporalities, or cultural presumptions about temporal process, hinge as much on the variously imagined ends of history as they do on its origins' (ibid.: 109). Here, a community's cosmology serves as a 'frame of reference for comprehending its present position within a historical trajectory of change' (ibid.).

The stories told by the Paiyamo did come together to create a history of the Project and people used them, to borrow Ballard's term, to position themselves within a historical trajectory of change from the days of the ancestors and *kastom* – when Afek hid *bangri* in what has since become the Project area – all the way to a future imagined around a vision of an operating mine and development at Frieda River. As I discuss in more detail elsewhere in this book, in the time I spent with them the Paiyamo rarely spoke of events that affected them as immediately important. They believed that the significance of any action or event would only ever be revealed to them over time, and they expected these revelations to take the form of a social relation, a material/physical alteration or a new opportunity for action (see chapter 3). Although the way in which they talked about their historical trajectory tended to resemble Gellner's model of neo-episodic time, they skilfully identified links between stories they told of the different epochs and moments of revelation in which they were able to see the significance or meaning of an event long after it took place. The stories themselves changed and evolved as the Paiyamo altered the way their past appeared and shifted the limits it posed on their current and future actions.

Studying this ability of stories and the acts of storytelling to restructure past events, changing the way we experience them and awarding them with new consequentiality, Michael Jackson (2002: 18) described storytelling as 'a coping strategy that involves making words stand for the world, and then, by manipulating them, changing one's experience of the world'. He argued that stories allow us to imagine and frame personal experiences in ways they can be shared, making them a tool for framing and managing social relations.[10] Crucially, for the context of this book, he described storytelling as a 'vital human strategy for sustaining a sense of agency in the face of disempowering circumstances' (ibid.: 15) and claimed that 'storytelling reworks and remodels subject-object relations in ways that subtly alter the balance between actor and acted upon, thus allowing us to feel that we actively participate in a world that for a moment seemed to discount, demean, and disempower us' (ibid.: 16). Many of the narratives that can be heard in Paupe during long evenings by the fire are not aimed at external audiences. Instead, they serve as a medium through which the Paiyamo seek to understand and explain to themselves various elements of the world they live in. But over the years, some of those stories have gained particular prominence and consequentiality in the broader context of the Frieda River Project and have become an important mechanism through which the Paiyamo partake in the process of planning for a mine at Frieda. Using a rich repertoire of stories, they embarked on a task of transforming their past into their history as the Project's stakeholders. However, as much as the Paiyamo stories and actions in the context of the Frieda River Project were driven by their desire and aspiration for development and the new opportunities this entailed, they could also be seen as part of an anti-politics machine that camouflaged serious power inequalities at play – a topic that I touched upon earlier and that will be explored in more detail in chapter 4.

Through the narratives they told of days past (which, to repeat after Jackson, are still in the making) the Paiyamo, their ancestors and their land became important agents awarding efficacy to the Frieda River Project. This ability to present their past in a particular way allowed the Paiyamo to expand the limits of their present and future and capitalise on the fact that, as argued by Hacking (1986: 167), 'for every person, in every era, the world is a plenitude of possibilities'. A version of this process has been noted in many parts of Papua New Guinea, where the expansion of the resource extraction sector 'had a remarkable effect on the production of indigenous narratives' (Jorgensen 2014: 23). For example, at Mount Kare, Wardlow (2004) recorded how, through a reformulation of myths, gold production was revealed as an ancient, ancestral undertaking. At Tolukuma gold mine, Hirsch (2004) described how, in the context of land rights, the mythical stories gained new significance, and previously public and widely known narratives became secret and more elaborate. At the fringes of the Kutubu Oil Project, Ernst (1999: 88) claimed that the newly emergent form of storytelling was a 'performance by all parties concerned, but it is messy, contested, and full of ambiguities, and it lacks the imperative clarity of liturgical language and performance. For these reasons storytelling can be a highly politicized discourse'. In his view, the political qualities of stories were located in their ability to formulate and reformulate relations. By telling a story, Ernst argued, the storyteller animates a transformative act, inevitably changing his or her relationship with their audience.

The ability of stories to transform relations between the storyteller(s) and their audience is crucial to understanding why the Paiyamo used stories to explain themselves to their network of social relations but also to explain the network itself. In this process, relations were being transformed, sometimes dramatically, as the Paiyamo used stories of past events to manage a range of internal and external relations, past, current and future, and to seek leverage in the face of structural inequalities present in the ongoing process of stakeholder engagement. The remainder of this chapter captures the ways in which the Paiyamo told me their story and the history of the Frieda River Project.

Paupe Village

By the time I travelled to Frieda, Paupe was an established village of some twenty-six households. According to the company census, 201 people lived there in 2012, 196 of whom declared themselves as belonging to the Paiyamo Amoke language group. Paupe was described to me as a Paiyamo community in which five different Paiyamo clans lived together. At the time, they were known as: Melefale, Nesiar, Suna, Inaku and Ayamo.[11] The sixth Paiyamo clan, the Imlou, lived in Duranmin. According to the local oral history, Paupe village (also known as Bawi, Bapi, Papi or Paupi) was established in 1962 by a kiap who arrived at Iniok from Ambunti station following the Sepik River. As it is remembered today, this arrival marks an important turning point in the history of the Paiyamo. In most conversations, and for the purpose of most stories, the arrival of the kiap and the founding of Paupe serve as a cut-off point at which the time of the ancestors ended and the time of the white people began.

Located in the East Sepik Province, a short boat ride away from the Sandaun provincial border, Paupe is a truly beautiful place, tucked away behind the vast waters of the mighty Sepik River and hidden from view by the thick, green overlay of a luscious, tropical rainforest. The village was accessible from the airstrip by a narrow, muddy and overgrown footpath, with the strong currents of the Frieda River to one side and the low hills stretched along the horizon to the other. Walking north, away from the airstrip, past the gate marking the boundary of the company Airstrip Camp and further past the old, dilapidated remains of what used to be a community clinic building, you would arrive at the first and southernmost of what are known locally as Paupe's three 'corners' (hamlets constituting part of the village). Named SBU Camp (or SBU, for short) after the Sepik Baptist Union, it was home to six Melefale households. The Melefale were the most numerous of the five Paiyamo groups in the community. Pastor Jacob was a Melefale man and the group's leader, as well as a spokesperson for all Paupe's Paiyamo. Although my research focused on Paupe and the Paiyamo, I spent most of my time and developed the closest relationship with the Melefale, with whom I lived in the SBU Camp. Most of the buildings were built from bush materials, and in the morning, narrow streaks of smoke could be seen above the houses as their inhabitants reignited last night's fires and prepared to start a new day. There were also, alongside the clinic, two other semi-permanent structures in SBU: a Baptist church with its faded green timber walls, concrete flooring and tinned roof; and a large, slowly deteriorating house with an iron clad roof, which was built by one of the companies conducting exploration work at Frieda and gifted to pastor Jacob and his family in 2004. This house became my home in Paupe. Pastor Jacob himself visited it only sporadically, preferring instead to live in Ok Isai or to stay in his garden house in Seligeme (some half an hour south on a motorised dugout canoe from Paupe). Pastor's daughter, Dinah, lived in the house with her daughter Mawi and took on the main responsibility for looking after me. Over time, the SBU Camp became my home and I got to know its ways and inhabitants well.

Every Sunday, the usually quiet SBU Camp became a hive of activity as community members from all over Paupe came together for a weekly Christian service or *lotu*, as they referred to it in Tok Pisin. Early in the morning, a couple of dugout canoes would arrive at a nearby waterfront with churchgoers from Aslimbum, the second of Paupe's three corners. Located directly across the rushing current of the Frieda River from SBU, Aslimbum was the smallest and newest part of Paupe. Four families had settled there in search of more easily accessible, fertile gardens. By far the largest group would emerge from the path leading into SBU from Aston, the main part of the village. Aston occupied a large clearing site some 20 minutes' walk north of the SBU Camp. Comprised of sixteen households, it also hosted on its grounds a number of village stores and a small building housing a company-provided CB radio used for communication across the Project area. Although a lot of the time the stores stood empty as availability of dry goods in the village was intermittent at best, they served as centres of social and economic activity and played an important role in a process through which the Paiyamo made claims of 'modernity' and prepared for development. Aston was also home to the community primary school with an adjacent teachers' house. The two buildings often stood empty as Paupe struggled to keep teachers delegated to the village from Ambunti, the District station.[12]

On the edge of Aston, local councillor Amos built his house. The title of village councillor is an elected local government role and, at the time, Amos was completing his fourth five-year term in office. Numerous wind houses[13] scattered around Aston offered inviting spaces for formal and informal community gatherings – places where both men and women got together to exchange stories, discuss current issues and get some rest on their way to or from the many gardens surrounding the village. Although small trade and alluvial gold mining as well as ad-hoc employment and financial arrangements made by the Paiyamo with the company developing the Project brought a relatively steady stream of cash into Paupe, the village had largely retained its subsistence-based character, and gardening, fishing and hunting continued to be primary occupations across the community. Aston was connected to the SBU Camp and further to the airstrip by an overgrown footpath tracing the western bank of the Frieda River. In the 1970s and 80s, the path was wide and served as an access road for company vehicles used in the early exploration activities at Frieda. But in the absence of cars, it narrowed significantly, and by the time I walked it for the first time it was just about navigable on foot.

Sunday church services at the SBU Camp were an integral part of Paupe's weekly rhythm. The services themselves lasted anything between 40 minutes and 2 hours and were led by pastor Jacob or, in his increasingly common absences, by one of the other Christian leaders living in the village. After the service was over, people tended to return to their houses with little haste; more often than not, long conversations, storytelling sessions and discussions on current affairs continued in the SBU Camp for some time after the church service had finished. It was during one sunny Sunday afternoon that, once the congregation had left the clearing outside the house I shared with Dinah and her daughter Mawi, pastor Jacob turned to me and proclaimed: 'Right. Coffee night. It is our *kastom* now. It is a night when we drink coffee and tell stories.' On that particular evening, there were seven of us in Dinah's large kitchen – myself, pastor Jacob, three men visiting Paupe from Duranmin, Dinah and her younger brother Aiyap. 'Some of those stories I would not tell to a woman or a child,' the pastor said as the water in an old kettle secured inconspicuously above the fire was coming to a boil. 'They are secret. They can only be heard by men in the cult house. If women or children hear them, the stories will kill them, poison them. But we will tell you, it's OK,' he added after a long pause.

This is how I began to hear the stories of the Paiyamo past, of the ways they lived and of the events that changed their world as they knew it, presenting them with new challenges and opportunities and ultimately leading to the emergence of new ways of living and new ways of being in a place. The stories were coming to me in different ways, settings and circumstances, from trivial, everyday exchanges over basic daily tasks to almost ritualistic moments where the setting, the build-up and the tone of voice of the storyteller indicated that an important moment was about to take place. At times, they were addressed to me alone, while on other occasions they were told in presence of other members of the community. Generally, storytelling was considered the domain of men, however, it was not uncommon for women and children to gather around and listen to the stories being told. On rare occasions, women would themselves take on the role of storyteller, bringing events and people to life through stories uttered in the gardens, kitchens or wind houses of Paupe.[14]

The *Kastom* Stayed on Tau

A long time ago, before white people arrived in the Frieda area and brought the Paiyamo speakers together, there were numerous distinctive groups scattered around, hidden deep in the bush. Each group held the rights to a distinctive area of land and had its own stories of origin. The groups traced their beginning to the time when their first ancestors arrived in the area bringing with them a range of skills, knowledge and powers. First ancestors were described as land spirits, known in Tok Pisin as *masalai*, who shared human and non-human characteristics and who could transform completely depending on need and circumstance. Each *masalai* had a real, true name that was only known to those men who were members of the local cult. They were commonly referred to in Tok Pisin as the *kastom lain*, and this is how I will refer to them here. The spirits also had public names, which were commonly known in the community and readily used in the stories the Paiyamo told about the time of their ancestors. Melefale and Imlou (the only Paiyamo group not residing in Paupe) traced their ancestry back to a crocodile they called Melesinapi (see chapter 4).[15] Together, they constituted a larger group called Yuupaiyamo Sane (described to me in terms of the Mountain Paiyamo or mimicking the terminology adopted by the neighbouring Miyanmin communities, the West Paiyamo). The remaining four groups living in Paupe were collectively referred to as the Suupaiyamo Sane (the River Paiyamo or the East Paiyamo): the Inaku claimed ancestry from a little bird with a red belly called Tobiasu; the Suna – from a white Suna tree; the Ayamo claimed to be descended from a limbum, a type of black palm used in house construction (known locally as Simao);[16] and the Nesiar from bamboo (which they referred to as Keywaime). By moving through the local geography and landscape, the spirits marked out land boundaries at Frieda. Descriptions of spirit travels – a living archive of Paiyamo land ownership – provided evidence of their claims to land in the Project area. The stories of individual groups' origins, together with the true names of the *masalai*, constituted an important source of power and identity among all Paupe's groups. The Paiyamo classified this form of narrative as particularly important and meaningful – tales that carried with them and within them great powers and that were ascribed almost magical qualities (see chapter 3). The *masalai* attributed each group with a range of powers, which they continue to use and which were still being revealed to them at the time of my research at Frieda (see chapters 3 and 4).

I did not find evidence of a shared origin story among all the various Paiyamo groups, although I have recorded an ancestral story linking Melefale and Suna, which was told to me on two separate occasions by the leaders of both groups. In Dinah's house, I also heard on numerous occasions the story of an old woman who was the ancestress of the Melefale and the Imlou. The narrative linked the Yuupaiyamo Sane ethnographically to the neighbouring Telefomin and further to other groups across the Mountain Ok region. A version of a closely related story was recorded by Roger Lohmann (2000: 98–101) during his field research in Duranmin. As it was told to me in Paupe in 2011, the story goes as follows:

> There was a woman who lived on Tau Mountain with her mother. In Paiyamo Amoke we call her *suia mebe* [old woman]. She was kind and gentle but her mother was anything but. The mother was filled with anger and violence. The two women were building a large house on Tau. One day, the mother asked her daughter to kill a man, drain his blood and use that blood to paint the posts supporting the house structure. But the daughter refused. The two women had a huge argument. In the end, they both left Tau. The mother headed towards Telefomin and eventually arrived at Telefolip. She brought with her war, violence, witchcraft and all different kinds of powerful magic. The Telefomin called her Afek.[17] The daughter travelled to a place where river Tau joins with river Pepe. Once there, she stopped for a meal. After she ate, her stomach swelled and she became pregnant. She continued her journeys through Wario, Uligeme, Yougene, and finally arrived at a mountain we call Usimi. There, she gave birth, but her child was not a human; it was a crystal-like stone called Wobsaseneso. This stone is still there. It continues to grow and crumble, and pieces of it can be found on the mountain slopes. Every time a piece breaks away, another one grows in its place. The Yuupaiyamo men sometimes go to Mount Usimi and carry a piece of the stone off the mountain. When in possession of the stone, they can access its many magical powers. The true name of the old woman is a taboo, a secret, but she has a public name which we use to tell stories about her. That name is Semiti.

Pastor Jacob was recognised throughout the Paupe community as a great and knowledgeable storyteller. 'His father, Aiyapanai was a very wise man. He knew all the stories and tabooed knowledge. Before he died, he gave them to Jacob, and now Jacob is the keeper of the stories' – this I would often hear all over Paupe. Aiyapanai himself has over the years become a legendary figure among the Paiyamo. He was famed for being an excellent warrior during an extensive period of tribal fighting that swept through the Frieda area in the first half of the twentieth century. After the war, as a local leader, he helped the kiap gather all the Paiyamo groups together into one village community and became its first *tultul* (local assistant to the colonial administration) when Paupe was established in 1962. His name is one of two featuring on the 1972 sale of land documentation described in chapter 1, and his role in the early encounters with the company as well as colonial administration is relatively well documented, his name featuring in a number of patrol reports from the 1970s and 80s. Possibly most of all, however, Aiyapanai is today remembered as a member of the *kastom lain*. Before Christianity arrived at Frieda, a welfare cult performed by men played an important role among groups that now define themselves as the Paiyamo. Men initiated into the cult had access to a body of restricted knowledge and participated in the *haus tambaran* (lit. the house of ancestral spirits, also used here to mean the cult house) *kastom*. In the cult house, they learned skills of dark magic and warfare, and how to secure *masalai's* help to ensure the wellbeing of the community. In the house, different aspects of the Paiyamo ways of being were produced and reproduced, awarding meaning and consequentiality to the actions of its people. The knowledge and power of the *haus tambaran* enabled the Paiyamo to live and be in a world full of uncertainty and ambiguity and to find ways to explain and engage with new events and situations.

All five Paiyamo groups in Paupe referred to *haus tambaran* when talking about their past. However, only one *haus tambaran* was remembered as having had a physical house-like structure. It was located some distance away from what is now Paupe, on the Tau Mountain, south-east of the Frieda Airstrip and near the Tau River, which sets the location for the ancestral myth of Semiti. It served as a cult house for the Melefale and Imlou before it was destroyed in 1977 during a Christian revival movement (see chapter 3). It was a rectangular structure, raised some two meters above the ground with a steep, angled roof. A fire used to burn in a clay hearth in the middle of the building, and bows, arrows, wild pig whiskers and string bags with the bones of male ancestors decorated its walls. The house was surrounded by a row of red tanget plants (cordyline fruticose, known in Paiyamo Amoke as *obiaro*) and red marita trees, and only men who belonged to the *kastom lain* could cross its threshold.[18] Members of Nesiar, Suna, Inaku and Ayamo acknowledged the existence of the cult house on Tau. They talked to me about temporary shacks they themselves would erect in the bush when the men needed to perform cult activities. The ways *haus tambaran* was described to me in Paupe spoke of a cult house as a physical but also intangible space that could be evoked when the older men wanted to pass knowledge and stories on to young initiates, or when an intervention of the cult's powers was required to solve a particular problem or address a specific concern facing the community. For all Paiyamo groups, boys would be inducted into the cult for the first time around the age of 16 to 18. There, alongside the art of sorcery, techniques of war, ancestral stories and the art of managing relationships with the *masalai*, they would learn practical skills of house building, hunting and growing crops (e.g. establishing taro gardens). It is almost certain that by that age the boys would know how to do those things anyway. But the true meaning of those activities, it was explained to me, could only be revealed to them by the *kastom lain* in the ritualistic space of a *haus tambaran* (see chapter 3).

It was the participation in the cult and the knowledge, experience and relations that were reproduced in the cult house that provided the foundations of Aiyapanai's power and authority. Although all men of the time would have been expected to take part in cult activities, most of the men of Aiyapanai's generation perished in a long period of intense tribal fighting immediately preceding the arrival of colonial administration at Frieda.[19] They took with them the secrets and stories of the *haus tambaran*, leaving few who were able to use its powers and pass its knowledge and skills on. It was a common sentiment among all Paupe's groups, who today can be heard saying that 'war, fight, *kastom* – it all stayed on Tau'. But although the Tau *haus tambaran* and all the people who crossed its threshold are now gone, the cult house continues to offer a conceptual and experiential space in which the Paiyamo men cultivate their knowledge of stories, spells and secrets, and which they use to make this knowledge meaningful and effective.

Fighting on the banks of the Frieda River involved all of the groups inhabiting the area and had a dramatic impact on its demography. During my stay in Paupe, there were very few men alive who remembered those events. One of them was Tikap.[20] Himself an Iwam speaker, he identified as a member of the Melefale group and lived in a house on top of a hill in the SBU Camp. He described the history of fighting at Frieda to me in the following words:

> In the time of our first ancestor, Obilibo Menaneme, there was no fighting here. Not yet. At the time of the second, the fighting had not yet started. It began at the time of our third ancestor, Grenabe. The Melefale fought with the Sanemaru, Yenayo, Orio and Tari. They fought over land, women, sorcery and theft. Back then, the Telefomin groups stayed far up the mountains, but they too came down and joined the fight against the Sanemaru. As a result, the Paiyamo and the Telefomin killed all the Sane-

> maru, who are no more. The fourth ancestor was Aiyapanai. In his time, big fighting came into the area. The Telefomin people attacked the Miyanmin, Sisimin, Oksapmin, Duranmin, Rebame, Sumare, Uriambo, Kuiaribo, Illigimin and the Paiyamo, as well as the Ountoumin. The fight started in Telefolip with the burning down of the Telefolip *haus tambaran*. In an act of revenge, the Telefomin killed all Illigimin, who are no more. Back then, Melefale and Ountoumin stayed at Melegeme. The Telefomin attacked the Paiyamo there, and the Ountoumin. One of the Ountoumin women was crossing a bridge. They call her Ameson. In her womb she was carrying a baby boy, Obrum (also known as Marisak). The Telefomin captured the woman and took her back with them. All the remaining Ountoumin were killed and are no more.[21]

The kind of narrative presented above is a popular one among the Paiyamo. It at once provides an account of local history and outlines the relationships between different groups, allowing them to lay claim to specific areas of ground – a role that has intensified significantly following the 1979 and 1996 landownership studies, which endorsed the re-created Ountoumin as principal landowners in the Frieda River Project area (see below). Although the Paiyamo managed to withstand the Telefomin attacks, they suffered great losses in the process. The population, which used to be high enough to allow for self-sufficiency of individual groups living in longhouse communities, was dramatically reduced and reached a critical low in the 1950s and early 60s.[22] While the group members say that they received great powers from the *masalai* at the time of fighting, the stories they tell paint the Telefomin as the more powerful, forbidding and dangerous opponents.[23] Making reference to the Yuupaiyamo Sane origin story recalled above, one of my informants in Paupe explained to me that the Telefomin drew their powers from the mother of the old woman, whom they called Afek. 'When the mother and daughter argued at Tau, the old woman took all of our powers and left. She took magic spells and the ways of war and brought them to Telefomin *haus tambaran* in Telefolip,' he told me. 'If only she did not take them, and if she had stayed with us, it would have been the Paiyamo who won the fight. It would have been us who attacked and conquered everyone. And now we would be the principal landowners for the company. If only.'[24]

Lineage Houses

All but one of the recognised Paiyamo groups lived in Paupe. They stayed in households that mimicked the Christian model of a satellite family unit, and almost all were made up of a kitchen and a separate house used for sleeping. The buildings were made of bush materials and required frequent repair. It was not uncommon for houses to be taken apart and new houses to be erected in their place. However, as I mentioned above, before a colonial administration patrol arrived on the shores of the Frieda River and brought the Paiyamo together into one village community, the different groups of Paiyamo speakers lived largely independently in longhouse communities dispersed around the broad Frieda area. These buildings were referred to by the Paiyamo as either *kastom* houses or as *lain* houses, where 'lain' can be translated roughly to 'lineage'.

The lineage houses were large, imposing structures built with bush materials and erected on land, skilfully and laboriously cleared from the dense rainforest. Although anthropological vernacular describes them as longhouses, a term widehouses might offer a more fitting description here, as the buildings were wide rather than long, their construction spreading sideways rather than front to back. The houses were raised high above the ground and accessible only through one narrow set of stairs carved from a single timber log propped against the building's long front wall. The stairs led onto a veranda which circled the structure's rectangular form. Four sets of doors led inside the house – two were used by men, the *kastom lain* (at the front and back of the building) and two by women and children (the side entrances). The space inside the house was strictly regimented, reflecting a series of taboos and reinforcing distinctive gender roles within the groups. The middle space, which surrounded a central pole that provided support for the high roof of the structure (*wan pauio* in Paiyamo Amoke), was the domain of men. The men tended to spend a significant proportion of their time in the bush. While in the house, they would eat, sleep, tell stories and pass time carving intricate designs onto bamboo arrows used for hunting. Spaces on the two far ends of the building belonged to women and children. There, around symmetrically located clay hearths (usually two or more on each side), they would spend time talking, cooking and making string bags, sheltering from the outside heat in the house's cool, spacious interior. Women were only ever allowed to enter the men's space for traditional *singsing* events, during which they were expected to either sit or stand at the very centre of the structure around the central pole. Life in a longhouse was a communal affair that did not leave much space for either privacy or intimacy. Numerous garden houses located up to two days' walk away from the main structure offered much needed respite and very welcome if often short-lived relief from collective food sharing obligations.

Before the period of tribal fighting left its mark on the area and its demography, the longhouse communities were largely self-sufficient. Although membership was loosely based on kinship structures and affiliation, the communities constituted fluid and open social systems in which adoption rights were as strong as kin ties and in which co-residency, shared experience and reciprocity links could over time create kin-like ties equal to these of blood relations. It was possible to marry within the longhouse

Illustration 2.2. Exterior of pastor Jacob's house in Seligeme. Photograph by the author.

community, although marriages outside longhouses were actively encouraged. Consequently, through marriage, the Paiyamo were connected to several other groups, most prominently Eliptamin, Duranmin, Telefomin, Wario and groups living in the May River area. Trade connections further embedded the longhouse communities in a dynamic regional network. The different Paiyamo longhouses were located a few days' walk away from each other, but the groups maintained regular contact and sustained strong ties, also through intermarriage and exchange. The communities moved around every few years in search of new, more fertile gardens and greater security. As Tikap once explained to me: 'In the time of unrest and fight, you need to move around. It you stay in one place for too long you will get killed.' The longhouse communities lived in peace with each other, and many care and reciprocity obligations extended beyond individual house lineages. Their members interacted frequently and often made the long and tiring journeys to visit one another. They assisted each other in a range of activities, such as house building, hunting and gardening, and they fought together against their enemies during local warfare. Every so often, they would gather together for indoor *singsing* events, where songs were sung until sunrise, telling stories of the different lineages and thus reminding people of their ancestry, landscape and sources of power (see chapter 4).

Longhouse communities dominated the Paiyamo spatial and social organisation for a long time. But they ended, like many other things, with the

Illustration 2.3. Interior of pastor Jacob's house in Seligeme. Photograph by the author.

arrival of colonial administration and the first Christian missionaries in the 1960s and early 1970s, with the last longhouses disappearing a decade later. In 2011–12, two structures resembling the big houses of days past could be found along the Frieda River: the old, big house perched on the top of the hill in SBU Camp, a description of which opened this chapter; and a much newer garden house in Seligeme, which pastor Jacob erected to, in his own words, 'teach the younger Paiyamo about their ancestry and *kastom*'. Despite the many years that have passed since the large longhouses were abandoned, they continue to serve as a source of distinctive identity for the five groups that form the Paupe community. The pattern of longhouse communities was still visible in Paupe, as most people claiming membership of a house lineage (Melefale, Nesiar, Suna, Inaku and Ayamo) erected their houses next to one another in the current settlement. However, the lineage houses themselves were regarded as a way of life that belongs firmly in the past, in the time of ancestors, *haus tambaran* and *kastom*.

The Time of Gold

By the late 1950s, the once self-sufficient and prosperous longhouse communities were struggling; the period of warfare had taken its toll and there were very few survivors. 'We weren't many,' one of my interlocutors remarked. 'Aiyapanai and his children, the Melefale group – that line had the highest number of survivors after the war. The second was Suna. Ayamo had one or two people left. Inaku were quite a few, but Nesiar only about three or five.'[25] The difficult, post-war reality provided a contextual backdrop for the arrival of the first colonial patrol in the area. Many of my informants in Paupe linked this arrival and the administration's presence in the area to truce and the end of tribal fighting. 'The government stopped the fighting and brought people together,' they told me on several occasions. In local narratives, this arrival marked the beginning of a new time, a new epoch for the Paiyamo. In 1962, twenty years after the German woman, Frieda, came to collect stone samples, a white man arrived by boat from the Sepik River and, like Frieda before him, was welcomed by Aiyapanai. 'The kiap gave dad a cap and said he was now the boss man, a *tultul*. He could see that dad was a chief here,' pastor Jacob told me one day over a rare can of Coca Cola. Together, the kiap and his newly appointed *tultul* (a Tok Pisin term for a village official and assistant to a village or tribal chief, a *luluai*, under colonial administration) gathered what remained of the different Paiyamo house lineages into a single administrative unit – Paupe (named after the village's original location on the eastern shore of the Frieda River). For the colonial administration, this process of community-making was part of a complex and widespread exercise in turning fluid and flexible social systems across the country into administrative units, thus making Papua New Guinea into a country that could be governed. But for the Paiyamo, it was a way of creating a new, opportunistic social structure. 'When the white man arrived in 1962, there were very few of us left. So when Aiyapanai came and said we should all move to one place on the Melefale land, we all moved,' I heard from a man identifying himself as a member of the Suna group. A Paiyamo-speaking community of Paupe was born.

The arrival of the first colonial patrol and the establishment of a Paupe village at Frieda were shortly followed by the arrival of the Carpentaria Exploration Company (CEC) in 1969. As described in the previous chapter, 'the company arrived in a helicopter and landed in Paupe'. The Paiyamo narratives speak about mixed reactions to CEC's arrival. On the one hand, they describe how most people run away to the bush, scared of the helicopter and worried about the return of war into the area. On the other hand, they depict people welcoming the company and rejoicing at the prospect of the work and money it would bring. Some of the narratives describing CEC's arrival contained within them a description of gifts and offerings received from the company in those initial encounters. Back then, it was a common practice and a matter of routine for colonial patrols, missionaries and commercial agents to use gifts such as salt, matches and rice as a means of negotiating early relationships with local communities all around Papua New Guinea. In many parts of the country, Frieda River included, these early processes of gifting and the obligation they created led to a particular kind of expectation and framed people's thinking about new opportunities these kinds of relationships could bring (leading, in some cases, to the emergence of a cult of cargo – see chapter 3). But as the gifts often changed hands before any other kind of relation was formed between local communities and the new arrivals, their true worth and cost were uncertain and largely unpredictable.[26] Many years later at Frieda, the Paiyamo could still be heard saying that the first company tricked them into accepting gifts that then implicated them in a series of opportunistic openings and obligations that they were not fully aware of at the time – and that were only now being revealed to them in their role as an impact community of the Frieda River Project.

Many stories of the first company's arrival focused on the role the Paiyamo played in the early days of what has over time become the Frieda River Project. Tikap once explained to me in great detail how he and a number of other men served as guides and porters, helping CEC set up its first camps in the area and making it possible for the company to conduct initial exploration work at Frieda. One of the oldest men in Paupe, Tikap was a *luluai* (a village chief) at the time of CEC's arrival. Characteristically of Paiyamo storytelling, Tikap's narrative was firmly embedded in the local landscape and described a series of journeys through the natural environment and a succession of marks that these journeys left behind. Tikap completed these journeys twice: once in 1969 with the company representatives and again in 1979 with a kiap called Tony Friend, who was tasked by the company to conduct a landowner identification exercise for the Frieda River Project. When he spoke to me all those years later, Tikap remarked that on none of those journeys had he encountered any Miyanmin or Telefomin communities: 'They were hiding. They only appeared later. This company and the Frieda River Project – it belongs to Paupe. It belongs to the Sepik. But now the Telefomin and the Miyanmin are claiming ownership of this ground. I am sorry, but where were they before.'

Another elder member of Paupe community told me that, one evening, having returned to Paupe from a day spent in the bush accompanying the company's geological team, Tikap turned to the Paiyamo and said: 'They have found gold and copper here. They will come back to get it – it will be a big job and it will take place here.' But the two words – gold and copper – now a prominent feature of the local daily discourse, meant very little to the community members fifty years ago. When the Paiyamo asked the company what gold and copper were, they were given no answer, no explanation. Instead, they say, the company gave them a hammer and told them to break the stone and see for themselves. 'When the white people came, it was a new thing,' I heard from a man whose father accompanied Aiyapanai during the early encounters with the mining company. 'At first, people did not understand why the white men were here. What gold was. They were probably told something by the company. But back then people did not understand; they just said yes.'[27] As time went by, the Paiyamo began to see gold and copper as a valuable resource; initially, at least, not so much because of its material value and the monetary wealth it entailed but because of the new opportunities they found in the relationships they formed with the many people who, over the years, travelled to this geographically remote corner of Papua New Guinea with the hope of finding their fortunes in the dense bush on the muddy banks of the Frieda River. Many years after they nervously welcomed CEC and first heard of gold and copper, the Paiyamo claimed that they were now no longer afraid of newcomers. For they came to see what they did not understand before: that it was the German woman, Frieda, who led the mining companies into the river valley (see the Prologue). 'She must have looked at the river and thought: I am sure there is gold here,' one of my informants told me. She did not say anything to the people at the time. Instead, the local story goes, she went back to Germany where she prepared a map of the Frieda area, marking on it the mineral deposits. Years later, CEC found this map and followed it to Paupe. 'After all' – members of the Paiyamo community explained to me – 'when the German woman was heading back towards the Sepik, the man from Iniok who accompanied her told us not to be scared if any more white people come'. And more white people did come – and they have kept coming (and going) since.

Looking for Landowners

By the time I began my research in Paupe, decades had passed since the German woman travelled up the Sepik River and the first colonial patrols and industry agents arrived into this beautiful and scarcely populated part of Papua New Guinea, exposing its inhabitants to a complex and widespread matrix of new kinds of internal politics and external influences. The longhouse communities, once scattered around, hidden from intrusive gaze in the depths of the rainforest, have disappeared, and the administrative village of Paupe has become home to a complex and fluid social entity – the Paiyamo Sane. The village has been moved a number of times to bring it closer to the newly constructed Frieda River Airstrip, for better access to company facilities and personnel, as well as ad-hoc employment opportunities. In the time I spent there, I came to see Paupe as a cosmopolitan community, the members of which were bound by flexible links emergent from their shared interests in being together and the potential this can create. I got to know them as being in-between places and relations, somewhere between the Sepik and the mountains, forming a community that emerged from and re-created for itself a culture that is incredibly open and creative. I began to understand that it is precisely that openness and creativity that enabled the Paiyamo to manoeuvre through the dynamic set of relations and relationships that the Frieda River Project has brought into being and to successfully perform its role as an impact community.

The term Paiyamo was well engrained in the Project vocabulary by the time of my research, although this had not always been the case. When in 1962 the administrative exercise in social control (undertaken eagerly by the colonial administration) marked Paupe on a map of Papua New Guinea, members of the different house lineages residing in the newly established village became known collectively as 'the people of Paupe' (see, for example, the sale of Piamoku documentation described in the previous chapter). Such declamation tied the village and its people to a particular time and place – at the far end of progress and modernity, on the remote and unruly banks of the Frieda River. When the sale of land contract was drawn up and signed on the northernmost tip of the newly erected airstrip in 1972, yet another label was ascribed to this newly established village community – one that termed them (and therefore made them) the owners of land at Frieda and, concurrently, landowners in the Frieda River Project area. Over time, this led to the emergence, in the mid-1990s, of a 'landowning clan' as a performative social category, enabling local communities to claim their right to a place in the process of planning for an operating mine at Frieda River. The Paiyamo clans took their names and drew their stories of origin from the old longhouse communities and the time of *kastom*. Like the house lineages, these newly emergent clans were complex and dynamic social systems in which cohabitation and cooperation could form the basis of kin-like relations and which contained within them a whole range of individual and communal experiences and aspirations.

Over the last three decades, categories of landowners and landowning clans have gained recognition and prominence across Papua New Guinea (see chapters 1 and 4). All over the country, they have come to constitute ideological idioms through which groups are able to lay claims against the state and the company for access to services, monetary royalties and compensation resulting from the presence of resource extraction projects in their area.[28] Nationwide, this emergence of landownership as a meaningful and influential legal category has led to a series of experiments with kinship through which groups and communities have competed among themselves for access to social, political and economic relations that such recognition entails.[29] In most places, these experiments have led to successful identification of landowning clans, satisfying the government and industry's desire for clan-based landholding units that can be actively engaged in the process of stimulating economic growth and raising rural incomes (Gilberthorpe and Banks 2012). However, it was a short-sighted victory for the companies as well as the state. All evidence suggests that 'the status of landowner in Papua New Guinea is, despite lawyerly desire for clarity, uncertain at best' (Jorgensen 1997: 599) as the process of entification defined by Ernst as 'making entities or things from what have been contingent categories' (1999: 89) is constantly challenged by local discourses of land and indigenous kinship and knowledge practices. As a result, what the government and developers are actually faced with is an ideology of landownership (Filer 1997a) acted out by 'the landowners', who often come to see it as 'a possibility from the perspective of a constitutive phenomenology of social life' (Ernst 2001: 126).

At Frieda, the transition from one still relatively new dominant form of recognition (an administrative village unit) to another (a landowning clan) was triggered by a series of land identification exercises ordered by the Papua New Guinean government and managed by the successive companies developing the Project. One of the early exercises in finding landowners in the Frieda area took place in 1979 and was conducted by Tony Friend, an Assistant District Officer based in Telefomin.[30] Friend visited three villages in the Project area: the Telefomin villages of Ok Isai and Wabia, and a Miyanmin village of Wameimin 2. He found that people living in those communities did not recognise land ownership boundaries and claimed that they were happy to share ownership of the territory under the Exploration Licence 58. He did advise, however, that 'if not for the present, then for the future, when perhaps other groups might make demands regarding ancient land rights . . . it is essential that, fairly quickly, a proper genealogy of all groups in the area is done by a qualified Anthropologist' (Friend 1979: 4). Consequently, a consultant anthropologist, Thomas Moylan, was engaged by the company and arrived at Frieda later that year. His task was to identify and describe the landowners in the Frieda River Project area and capture their affiliations with other groups.

Moylan generously shared his report with me. The document arrived on my desk one rainy morning in a large, thick envelope. It captured in great detail the complex history of the area; the war and conquest that began in the late 1800s and lasted until the early 1960s, when the first colonial patrols and missionaries reached this part of New Guinea. Moylan scrupulously fulfilled his task and recorded, as advised by Friend and contracted by the company, extensive genealogies at Wameimin, Wabia and Ok Isai. Through his description, the three communities emerged as socially coherent but greatly diverse groups containing numerous peoples who originally belonged to different cohorts but who were captured during the period of tribal fighting. By 1979, these captives were claimed to have been fully incorporated into their new communities and to have lost all rights in their home groups. The report offered recommendations on how compensation should be distributed in Wabia, Ok Isai and Wameimin. But it also warned about the complexities of land tenure and land rights in the area. Critically, it named the by then extinct Ountoumin as the original owners of land under EL-58, alongside the 'Paupe people of lower Frieda' [*sic*] and Wario people. As a result of what Moylan described as a genocidal war of conquest, 'the Ountoumin clan was reduced to effective extinction while the Paupe and Wario-Inkiapmin people were pushed to the north and east respectively' (1979: 1). At the same time, Ok Isai and Wabia enlarged their populations by taking captives from the conquered groups. Among those captives was a pregnant Ountoumin woman, who was brought to the village by one of the Ok Isai founders, a Telefomin man called Maisep. Recorded by Moylan in 1979, this story has since become one of the most effective but also one of the most contentious narratives in the process of landowner identification for the Frieda River Project.

'At some future time', Moylan (1979: 16–17) wrote in his report 'Paupe people may reassert themselves and press claims for lost lands. It is worth noting that almost all of the names of rivers and other geographic features are not of Telefomin origin but more likely Paupe or Wameimin origin'. Similarly to Tony Friend before him, Moylan concluded that 'further land boundary disputes can be anticipated' and pointed out that in the context of Frieda's deposits 'the whole notion of "traditional" claims in this area is misleading, all Telefomin claims in this area are recent and derive from their having exterminated the previous owners' (ibid.: 16). The remarks made by both men early on in the landowner identification process suggest that the category of landowning clans has from the start been problematic at Frieda, where ambiguity of social forms, combined with a long and dramatic history of conquest and occupation, formed a socio-political system that stubbornly resisted administrative efforts to create locally a collective social form based on ancestral, landowning clans that could lend itself to a system of bureaucratic control. To quote Jorgensen (1997: 620), identifying the landowners in the Frieda River Project area has been from its early days 'an exercise in alignment and self-definition in which the calculus of advantage intersects with questions of identity, all of which is played out against a backdrop of regional political relations'.

Following the two reports, the language of landownership and landowning clans entered the local lexicon not only in the Project meeting rooms and corporate boardrooms but also in the wind houses and open fire kitchens in local communities. By the time the next major land identification exercise took place in the mid-1990s, the ideology of landownership had gained significantly in political prominence, as local people were more knowledgeable about the efficacy of the landowner category. In 1995, unsuccessful attempts to reconcile Miyanmin and Telefomin claims to land led to 'a coalition of people claiming descent from the aboriginal inhabitants [who] have declared their intention of seeking registration as landowners as members of the Ountoumin clan' (Jorgensen 1997: 621; a process described in more detail in Jorgensen 2007a and Gardner 2004). Informed by their earlier experiences in asserting and evidencing membership of particular groups, they based their claims on historical narratives of genealogical descent from the Ountoumin taken captive in warfare. But, as Jorgensen (1997) notes, this process ignored adoptive kin relations and marked a departure from the local custom and forms of social organisation. The Paiyamo told me that at the time they spoke out and claimed that they, and not the Telefomin or Miyanmin groups, were the nearest living relatives of the Ountoumin, but the court dismissed their claims. In 1996, it ruled the reinstated Ountoumin (some of whom lived in Telefomin, some in Miyanmin impact villages) as the primary landowners in the Frieda River Project area. Far from settling matters, this decision led to a land dispute that continued to unravel at Frieda as the malleable and fluid relations and interests continued to compete with each other and as people's complex associations with land resisted the confines of the imposed administrative categories (see chapter 4).

In the time I spent with them, the Paiyamo attributed great qualities and powers to the label of landowner and saw it as a mechanism through which relations were both created and reflected, leading to openings through which access to the Frieda River Project's benefits could be obtained – whatever these benefits might be. Although their 1996 claim based on their ancestral relationship with the Ountoumin did not bring the desired outcome, the event offered a series of other opportunities, seized by the Paiyamo with inquisitiveness and playful enthusiasm. Seeing what claims to a particular identity presented in a particular form can do, they embraced clan terminology and willingly presented an image of a Paiyamo community within which distinctive ancestral landowning clans could be identified (see chapter 4). As the leader of the Inaku clan told me:

> When the time came for the land mediation[31] they were all looking for clans. They asked us all – now, which clan do you belong to? That was in 1996. Before then, we had the house lineage names and we would only use those names. We would not call ourselves a clan. Or a landowner. These words come from a white man's Tok Pisin. But we do now.

The emergence of the Paiyamo 'clans', therefore, supported Jorgensen's (2007a: 69) argument that 'it will always prove possible to find clans (or other such groups) if one tries hard enough, particularly if local people have a stake in making this possible [and even if] such entities might prove less stable and substantial than government officials (or mining executives) might like'. The Paiyamo readily talked to me about their clans in the context of the Frieda River Project. But it did not take much time or great observatory skills to realise that they used the clan terminology to assert and communicate not an ancestral kinship system of land ownership envisaged by the state and industry actors but a much more contemporary and dynamic political entity that manifested their shared interest in mining development at Frieda and willingness to engage with the company and government representatives at the Project. The words 'clan' and 'landowner' were commonly used in Paupe in the context of the resource development, and it was the Frieda River Project itself that awarded these terms consequentiality and made them meaningful. When not speaking about the Project, the Paiyamo used other labels, describing different affiliations within the community in terms of groups and house lineages or, when speaking about individual community members, in terms of kinship relations. It is important to note here that the idea of a landowning clan was neither trivialised nor externalised by the Paiyamo. Quite the opposite: it was awarded social consequentiality – individual as well as collective. It became not just a type of outwards-facing performance, the aim of which was to present a group of people in a particular form to a particular audience, but an important factor altering and shaping the world of the Paiyamo as they knew and experienced it, as I describe in more detail in chapter 4.

Conclusions

As landowners and, what follows, stakeholders of one of the biggest prospective resource extraction developments in the Pacific region, the Paiyamo found themselves caught up in a world of unprecedented interconnectedness and interdependency. Although in many ways geographically remote, surrounded by the vast waters of the mighty Sepik River to the north and ridges of the New Guinea Central Range to the south, they were fully and unpityingly exposed to a range of powerful external influences and forces reaching far beyond the river valley. Not all of these forces were visible to them – not all of them were known – but they had a direct effect on their lives and fed into the ways they imagined their future. Through a network of extended social, political and economic interdependencies, the Paiyamo were party to a whole range of dynamic relations with company shareholders as well as the Project's stakeholders, including national governments, international organisations and markets, financial bodies and networks of accountability. These systems contributed in one way or another to the process of 'making up the Paiyamo' – alongside the ancestors, the company, the government, and other agents who entered their world. The term Paiyamo itself, although commonly used by my hosts as a form of description and a statement of affiliation, emerged as a result of dynamic events and relations, some of which were described above.

The arrival of colonial administration and mining companies in the area introduced both opportunities and incentives for the local people to make their relations appear in a particular form and, Hacking and Hirsch would argue, to become of a particular kind. It seems quite ironic that in order to be recognised and able to exercise leverage in that fluid network of global movement and transnational flows the Paiyamo were expected to mould themselves into a kind of fixed social form that tied them to a clearly defined and finite area of land and branded them its owners. As chapter 4 will discuss in more detail, the Paiyamo embraced a landowner label which made them of a new kind, but only insofar as 'every practical change is also a cultural reproduction' (Sahlins 1985: 144). They skilfully aligned into ancestral, landowning clans that the company and the government could 'find' and categorise as stakeholders of the Frieda River Project. But the shape and duration of these clan forms were entirely caught up with locally negotiated and externally influenced factors and awarded meaning by a system of structural (legal, economic and political) and social relations that in themselves were constantly evolving and adjusting. This made 'making up kinds' with regards to the Frieda River Project an ongoing and deeply relational process.

The events presented here are seemingly ordered in the act of storytelling into different epochs punctuated by times of great transition. These were often dramatic but never complete in a way stipulated by Gellner, each revealing novel meaning and capacities in past stories and events. Embedded in local experiences of time – where no event can ever be immediately knowable and where to an extent, at least, the future has already happened but has not yet revealed itself (see chapter 4) – local histories of the Frieda River Project were constantly in the making. The previous chapter used company documentation to trace the history of the Project back to the late 1960s, when the first deposits were formally recorded, exploration licence issued, and when the CEC arrived at Frieda to begin assessing the area's mineralisation. Shifting perspective, this chapter followed the Paiyamo in tracing the history of the Project back to the time of ancestors and *kastom*, when Afek hid *bangri* in the Project area and the *masalai* marked the land and beginnings of different Paiyamo groups at Frieda. As will be discussed in more detail in the latter chapters of this book, this local understanding of time and Project temporalities meant that while the company and the government conceptualised the mine at Frieda as a future project, for the Paiyamo the mine had begun at the time of the ancestors and was already happening – a form of volition 'that makes things happen, but is not always visible' (Lohmann 2003: 183).

Notes

1. The Paiyamo used the term *Sane* in reference to groups of people: Paiyamo Sane, Melefale Sane, etc.
2. Wagner (1986) argued that we tend to measure time through its passage and not its presence. The happening of time, he claimed, can only ever be experienced in terms of an incident or an event or, as he chose to call it, an 'epoch'. 'This word', he wrote, 'is often used in contrast to the "interval" of time measurement . . . "Interval" is, after all, the very essence of using space to represent time, and an interval is measured by some kind of motion across a spatial traverse. The "stoppage" of epoch stands in contradistinction to this motion, for it constitutes a self-defined "piece" of time, something that is original and unmeasurable beyond all attempts at conventionalizing. Epoch, then, is time considered as organic, happening as one and the same as the frame within which it is perceived' (1986: 85).
3. While 'episodic time' entails a single transition between two states, neo-episodic time entails multiple transitions between a number of states (see Gellner 1964). For example, following his field research in the country's East New Britain Province, Frederick Errington described how, in the world of the Karavaran people, 'social change takes place through decisive leaps rather than through gradual development, and such changes are not implicit in existing events' (1974: 257). Among the Karavaran, he argued, social change is not conceived of as a gradual process, taking instead a form of decisive events entailing abrupt transition between one state of being and another. Like my research among the Paiyamo, his study of the Karavaran led him to argue that the decisive leaps that tend to characterise social change in the community require intervention of outside forces, albeit for a very different reason. Wolfgang Kempf's study of Christianisation among the Sibog in the Madang Province offers another noteworthy account of neo-episodic temporality in Papua New Guinea. In his work, Kempf wrote about the influence of the neo-episodic concept of change on Sibog behaviour patterns and argued that 'by adopting Christianity, the Sibog hoped to bring about a rapid and deep-seated change' (1992: 80). This hope, he argued, could only be fully realised if the people renounced everything from the previous phase, as complete discontinuity was an essential prerequisite for the new change occurring.
4. This comment echoes some of the themes that emerged from a number of ethnographic studies in the region; for example, Thomas Ernst's work with the Onabasulu (Southern Highlands Province), where, Ernst observed, exploration activities and the state's desire for codification and categorisation of collective life led to the emergence of 'a *purposeful* reality to Onabasulu "being" that [was] entirely novel, a function of contemporary circumstances and a manipulable state power, which exerts itself in the context of international resource

development and the need of multinationals to identify interest groups and concerned parties' (1999: 96, emphasis added). Although the new descriptions created novel ways of being for the Onabasulu, Ernst argued, these were 'based in understandings and concepts that are specifically and exclusively a part of the region, and may well have been previously mobilized in relation, say, to earlier Huli ritual expansion' (ibid.).

5. In his response to Stewart and Strathern (2002), Hirsch (2002: 370) defended the historical distinctiveness of the notion of 'human kinds' and its local implications and argued that, as conceptualised by Hacking, it is 'of fairly recent origin, arising out of the amalgamation of the enumerating state and the application of quantitative and qualitative reasoning to issues of human behaviour; (scientific) laws precise enough to predict individual action'.
6. Hacking (1995a: 368) himself spoke of the ability of kinds to 'enable us to redescribe our past to the extent that people come to experience new past'.
7. I do not want to suggest here that *all* such transformations were calculated and strategic. By their very nature, stories in oral traditions change over time – memory and repetition alters them and shifting contexts emphasise some parts of the stories while neglecting others.
8. Ballard (2014: 100) uses the term 'historicity' to refer to a broad field of anthropological enquiry 'into culturally specific forms of consciousness of the past'.
9. See Michael Goddard's (2013) work, where he analyses three histories of Duogo Island, Papua New Guinea, to highlight the importance of storytelling and the creativity of local historicity.
10. Hannah Arendt (1958: 97), whose writing inspired Jackson's book, argued that stories serve as a strategy people utilise to make private messages public: 'Action and speech are indeed two activities whose end result will always be a story with enough coherence to be told, no matter how accidental or haphazard the single events and their causation may appear to be'.
11. I used the term 'at the time' here as those names carry within them their own temporalities. In 1984, when Dan Jorgensen visited Paupe, the different groups he encountered called themselves Imaka, Koyali, Malipale, Pwimuku and Wasu.
12. Two teachers arrived in Paupe around the time I began my fieldwork. They were trained and employed by the government, although the company took on a large share of responsibility regarding the teachers' transport, provision of building materials for the school and the teachers' house, stationery supplies and banking assistance. The first of the two teachers fell ill shortly after arriving in Paupe and was evacuated to her parents' home at Ambunti station. The second teacher took on all the school duties but resigned after a few months, quoting the remoteness of the area and the lack of community support as key reasons for leaving. Although the Paupe community were disappointed by the loss of a yet another teacher, they felt that it was the company's responsibility and not theirs to provide food and essentials for the teaching staff. They accepted the

teacher's departure as a yet another sign of the company not fulfilling its social responsibilities.

13. Raised, roofed platforms made from bush materials.
14. The stories recorded here were told to me in Tok Pisin. Although the local language, Paiyamo Amoke, was commonly used in conversations between community members, Tok Pisin dominated daily communication and was the principal language used in interaction between different stakeholder groups of the Frieda River Project. Rarely would the young children in the village speak the local language, even though they understood it, preferring instead to answer any questions in Pidgin. When I asked my informants about words and turns of phrases in the local vernacular they would often refer me to older members of the community, saying: 'We do not really know deep inside Paiyamo Amoke.' Regrettably, I did not gain sufficient knowledge of Paiyamo Amoke to hold conversations and collect ethnographic material in the local vernacular. Instead, I conducted my research in Tok Pisin.
15. The Imlou lived in Duranmin, a community located a fair distance South from the Frieda Airstrip and not easily accessible from Paupe. They are better known as the Asabano of Duranmin (see Lohmann 2000).
16. The Jacaranda dictionary of Tok Pisin translates limbum as Kentiopsis Archontophoenix.
17. See *Children of Afek* by Craig and Hyndman (1990).
18. In one account I came across during my fieldwork, the big *haus tambaran* (*nuiami* in Paiyamo Amoke) was accompanied by a smaller structure (*nosuami*). In that account, the big house served the cult of warfare while the smaller structure served the cult of taro.
19. Aiyapanai himself died in 1989.
20. Tikap died in his house in Paupe in 2014.
21. In the local storytelling, this marks the beginning of the myth of the Ountoumin. The myth provided the foundations for a land dispute between the Telefomin and the Miyanmin, which began in the early 1990s and which was continuing to unfold at the time of my fieldwork (see also Jorgensen 2007a).
22. Demographic effects of war were likely exacerbated by influenza, which was present in the area at the time, although the lack of patrol documents from Frieda means that this is an informed assumption and not a statement of fact.
23. The Paiyamo say that their *haus tambaran* contained many useful spells as well as an extensive knowledge of magic that could be used in the battlefield. Ayapanai himself is said to have been the owner of a magical arrow (which he could hide behind and become invisible).
24. The same sentiment was recorded by Roger Lohmann (2000) among the Asabano in Duranmin.
25. The 1972 sale of land documentation recorded twenty-one names of people living in Paupe: sixteen 'adults' and five 'minors'.

26. Across the Highlands, Hirsch (2001a) recorded stories told among the Fuyuge-speaking communities in Central Province, where the white men were faced with the task of presenting a gift that was of interest to the local communities. He described how 'the highlanders were waiting for the white men to reveal forms that were analogous to and substitutable for what they already had. Once the white men had revealed these valuables, the highlanders were compelled to act in return' (Hirsch 2001a: 243), coerced into a particular kind of relation.
27. This statement reflects a popular argument used by the Paiyamo to undermine early arrangements made between the local groups and new arrivals, which they no longer saw as beneficial.
28. See, for example, Filer (1997a, 1997b, 1999) and Gilberthorpe (2007).
29. See, for example, Golub (2006), Sagir (2004) and Weiner (1998a, 1998b).
30. Two other official land investigation exercises preceded Friend's patrol. In 1969, B.M. Fisher, Assistant District Officer at Telefomin, was asked to delineate the land boundaries in EL-58. Fisher completed his patrol in early February 1970. In 1978, the second land investigation was conducted by a government officer, M. Kaalund. Neither of them reached Paupe.
31. The Paiyamo tend to refer to any exercise aimed at identifying land ownership structures in the Project area as 'land mediation' rather than 'landowner identification'. In other contexts, 'landowner mediation' describes a process of mediating between individuals and groups with contesting claims to areas of land.

Illustration 3.1. Dinah returning home from her garden. Photograph by the author.

3

Make-Believe

Introduction

'You see', it was explained to me one rainy afternoon,

> where there is rain, there should also be thunder. That means that the rain is just a normal rain, and that kind of rain is good for the gardens. If it is raining but there is no thunder – this means that the rain was caused by people. For example, if I am upset with my uncle, I can cause the rain to fall. Young people like me are not trusted not to cause rain unnecessarily so we do not know this magic – but I can ask one of the elders and they can do it for me. He takes the red tanget leaf, hides it under a stone in the river and utters a spell. As long as the leaf stays there, the rain will continue to fall. If it rains, my uncle will not be able to go to a garden but neither will I. Too much rain can cause a landslide, which may potentially destroy all gardens in the area. So when the rain does not stop, people try to find out who caused the rain and why. Once they know it is me, they confront me and try to resolve the issue. My uncle has to apologise to me, and I too have to apologise to my uncle. When the upset is over, the elder man removes the leaf from the water, makes the rain stop and we can go to the garden again. The spells belong to *kastom* and in the past the men would learn this kind of magic in *haus tambaran*. But now we are Christian and we know God. Very few people still remember the magic and the spells, and those who do are reluctant to teach those skills to others.

In the time I spent in Paupe, I heard many stories like this one. It seemed to be common knowledge among the people I lived with that incidents such as unusual rain causing damage to the garden harvest, rats spoiling a stock of rice in a village shop, or a wild pig charging at a person as they walk along a path marked out in the bush were, more often than not, signs of some form of wrongdoing, conflict or negative tension within the community that needed to be resolved.[1]

This chapter uses local ideas of personhood and knowledge to describe ways in which the Paiyamo engaged with supernatural forces in their world, where supernatural as an etic category is modelled as 'a volition that makes things happen, but is not always visible' (Lohmann 2003: 183).[2] For the Paiyamo, as with many Melanesian societies, the 'supernatural' is not separate from the 'natural'. It entails a hidden realm in constant interplay with the visible realm that can be elicited and directed through ritualistic activity, enacted beliefs and managing a wide range of relations. To that end, the category of 'supernatural' can be applied to the *masalai* as well as the Christian God. What I propose here is that it can also be applied to the mine at Frieda River, which is not yet visible but which, from the perspective of the Paiyamo, has already started to show effect (see chapter 5). The chapter explores religious lives of the Paiyamo and the religious practices they used to enact their beliefs. It uncovers a form of engagement made possible by local ideas about personhoods, knowledge and social relations. In this chapter I continue to explore the themes of temporality and change. Earlier, I challenged Gellner's (1964) theorising about episodic temporalities that frame Melanesian people's experiences and movement through time in terms of continuous, static periods separated by events of complete transformation. Such an episodic view of time fails to accommodate the life processes of the people it sets out to describe placing them, to borrow from Tim Ingold (2016: 130), in limbo, 'perilously suspended between two different worlds and two different times to which – more often than not – are affixed the labels "traditional" and "modern" respectively'. The events described below speak of periods characterised by the uneven pace of change across diverse aspects of people's lives. Neither denying nor trivialising the great importance and magnitude of distinctive transformative events, my approach circumnavigates the temptation of romanticising the Paiyamo, representing them as either unchanging and ahistorical or capable only of total and dramatic transformations grounded in discontinuity between one state of being and the next. Instead, I address questions of agency and power, both personal and communal. The chapter does not seek to establish what changes and what does not. Instead, it explores factors affecting conditions for change and the ways the Paiyamo try to control and direct the changes that affect them.

The Melanesian (In)Dividual

About half way through my fieldwork, I fell ill. There is never a good time for illness in the field, but the Saturday when I woke up feeling unwell was particularly poorly timed. A football league in Ok Isai was coming to the end of the season, and my SBU Camp family, together with a number of other people from Paupe, were planning a trip up the river to watch the season's final games and to participate in the jubilant excitement that the event provoked in all three communities along the Frieda River. Not wanting to spoil the day for everyone else, I decided to join Dinah and Mawi and nested myself uneasily in the dugout canoe as it fought against the strong currents. When we got there, the Ok Isai waterfront was buzzing with excitement and activity. From there, we took a short, twenty-minute walk inland, where, at one of the village clearings, two temporary gates had been constructed from lean sticks, forming a sizeable if somewhat uneven football pitch which meant that one of the teams was forced to attack uphill. Around the pitch, people from all three communities gathered to watch the games and pass time gossiping, exchanging stories and indulging in market food items prepared and sold by the local women. Although altogether it was a very enjoyable day for everyone involved, I struggled. The illness was making me weak, and by the time the games ended mid-afternoon I felt tired and grumpy and ready to head back home. We did not leave Ok Isai until much later that day, and the community youth group and a couple of district health workers from Telefomin decided to travel with us to Paupe to attend the Sunday church service and to conduct a women's immunisation programme the following day.[3] Shortly after our canoes left the Ok Isai waterfront, a decision was made not to head all the way down to Paupe that day but to sleep instead in pastor Jacob's house in Seligeme, about half way between the two villages. That night, the pastor's large house was filled with people, many of whom I did not know. I was not in good form that evening. The fever and the dizziness were taking over my body, and I fell asleep against the wall of the house shortly after I sat down. Not long afterwards, Dinah woke me up and helped me over to one of the four fires lit in the house. I lay down by the fire and stayed there, as Dinah later told me, like an old person who is weak and cold.

I tried to fall asleep and get some rest but kept thinking about the carefully sealed bag of medications I had left at the SBU Camp, not expecting to be spending the night away from home. I could feel Dinah's hand touching my arm. She stroked it gently for a little while and then put her palm on my forearm and kept it there. At different points through the night, I could feel people touching my skin: all I wanted that night was for people to leave me alone. I was annoyed at the constant touching but too tired to ask them to stop. In the morning, I felt better, relieved at the imminent prospect of heading home. Around me, my family and their guests were preparing to leave Seligeme and head down to Paupe. Before we left the house, Dinah passed me a boiled sweet potato, which I shared with Mawi when we got into the canoe. That evening as we were sitting by the fire back at SBU Camp and talking about the immunisation programme, Dinah's birth mother, Lucy, turned to me and started naming the people who had touched my forearm the night before. Trying to hide my annoyance with the night's disturbances, I expressed my gratitude for the way they had looked after me. It was not until later that night that she told me that when a person is ill some negative force takes over their body; the person needs to be moved next to a fire and kept warm, but sometimes the force is too big for the body to deal with it on its own, which is why people touch the skin of the ill person – to take away little bits of the illness. This way, Lucy told me, the illness is broken up into smaller pieces and shared by many people. That night, in Jacob's large house in Seligeme, the people touching my skin took some of the illness away from me and onto themselves, helping me to get better. Quite literally, they took the ill parts of me away and replaced them with parts of their stronger and healthier selves.

This image brings to mind Marilyn Strathern's (1988, 1991, 1999) work on dividuality of Melanesian personhoods, where a dividual person is made up of detachments of others, which he or she receives through elicitive, reciprocal exchanges. In a similar vein, Roy Wagner described a 'fractal person'. Much like Strathern's 'dividual', Wagner's (1991: 163) 'fractal person' is neither singular nor plural, 'never a unit standing in relation to an aggregate, or an aggregate standing in relation to a unit, but always an entity with relationship integrally implied'.[4] Both these models highlight the importance of relationality in Melanesian societies and speak of transformative qualities of relations. Strathern (2004: 102) further elaborated on the form of those relations and argued that 'in making connections visible, people assert their ever-present capacity to act upon them'. It is possible to apply this argument to stakeholder categories and the ways they made visible connections and relations between people in order to acquire new capacities in the context of the Frieda River Project. In particular, it can be used to interpret the ways in which the Paiyamo formed 'landowning clans' by demonstrating and making visible their connections to each other and their land to assert what they saw as an opportunistic position in relation to the Project and its other stakeholders (see chapter 4).

In her work, to demonstrate the difference perspective makes, Strathern contrasted the Melanesian 'dividual' with the 'Western' ideology of an 'individual'.[5] A number of responses to Strathern from the field of anthropology challenge this distinction. They argued that 'the modern self is every bit as fluid and dynamic as the (supposedly) traditional dividuals in Melanesia' (Smith 2012: 55); that 'the ideology of the Western person as fully individual only partially conceals the reality that Western persons are interdependent, defined in relation to others, depend on others for knowledge about themselves, grasp power as the ability to do and act, grow as the beneficiary of others' actions, and so forth' (LiPuma 1998: 60); and that Western and

Melanesian cultures 'both distinguish between the person as a uniquely constituted being and the person as a relational, social being. In both situations, the contexts in which individuation is deemed significant are often very different from those where social integration is paramount' (Macintyre 1995: 31–32). The consensus and the starting point of more recent debates is that people, be they Western, Melanesian or other, are both dividuals and individuals (Englund and Leach 2000) – in that in one way or another all personhoods are relational. LiPuma (1998: 57) argued further that 'persons emerge precisely from that tension between dividual and individual aspects/relations'. He conceived of the conditions of this tension as socially and culturally variable and proposed that the new ethnographic goal 'is to uncover the conditions . . . under which dividual and individual aspects of personhood emerge and are hidden' (ibid.: 61).

These arguments have guided much of the recent anthropological writing on the subject of the religious lives of people in Melanesia, especially in the context of their conversion from traditional religions to Christianity.[6] There is a sizeable amount of literature that looks at the premise of the 'individualising' influence of the latter, pointing to the shifting ideas of agency and responsibility and linking this influence to the creation of conditions for a capitalist economy, thus laying the groundwork for the process of economic and political modernisation (e.g. Bialecki, Haynes and Robins 2008). There is possibly even more literature that speaks to the individualising character of market capitalism and modernity itself (see, e.g., Douglas and Isherwood 1979; Turner 1988). Over the past fifty years, mining and Christianity have become fundamental to the ways in which my hosts in Paupe engage with and claim participation in the projects of development and modernity.[7] There can be little doubt that Christianity and the mining industry's interests and presence in the area introduced new ways for the Paiyamo to be, challenging a range of local ideas – including those about personhood, agency, responsibility and relations. But the dramatic changes experienced by the Paiyamo have also encouraged them to re-evaluate and in some cases re-emphasise some of the long-standing local ideas akin to the relational model proposed by Strathern, especially in respect to people's capacity for diverse relations and for using relations as a vehicle for personal and communal transformation. While some situations such as engaging in paid contract work with the company endorsed individuality, people in Paupe were often reminded of their communal obligations and the wide range of relations that, Strathern might argue, produced them (as chapter 5 will show).

The Paiyamo themselves did not reflect on whether they were individual or dividual persons. But they did reflect on whether they were good people and what being a good person means in different contexts and circumstances. They were aware of the fact that their actions had consequences, and they knew the effects of these actions (both positive and negative) could show themselves in other people, the landscape or in events and happenings – as illustrated in the story opening this chapter.

Return to the Haus Tambaran

Before they 'heard the word of God' and became Christian, Paiyamo religious beliefs centred around a cult performed by men that was inherently tied to the concealment and strategic revelation of religious knowledge. It was not uncommon to hear members of the community say that in the time of *kastom* and ancestors terrible things would happen should this knowledge be revealed to someone outside the *haus tambaran*. In some instances, even the act of discussing important stories and religious knowledge unnecessarily, regardless of context, was a dangerous activity. For example, Melefale often spoke of Melesinapi, a crocodile from Nena junction (see below). Stories of a two-headed snake, the second of the group's most important *masalai*, were uttered much less frequently. They would explain that while the crocodile is generally calm and can be talked about freely (as long as his name remains hidden), the snake is known for his anger and bad temper. 'If people talk about him too much, rain will come and the waters of the Frieda River will turn into a sea – drowning gardens and destroying villages,' they told me. In the *haus tambaran*, access to religious knowledge had to be carefully managed for the doctrine to be effective in sustaining the group, ensuring their wellbeing and providing skills and powers needed to fight their enemies and defend themselves in times of war. The management of all sacred knowledge was the responsibility of the cult and cult leaders, whose role was to keep the powerful knowledge hidden from the public domain and also maintain it through tactically passing the stories and secrets on to the next generations of Paiyamo men.

The men entered the *haus tambaran* for the first time as initiates 'around number five shave', at the age of 16–18. Once inside, they were able to eat foods forbidden to women, children and uninitiated men (such as red sugarcane, wild pig, yellow and red marita [*Pandanus conoideus*] and certain types of fish), see sacred objects (such as the bones of ancestral warriors) and hear powerful stories spoken by their fathers and grandfathers. The stories told in the *haus tambaran* explained the origins of the different groups in the area. They were sources of knowledge about natural and supernatural phenomena and elucidated many aspects of people's lives such as kinship, landscape, illness and misfortune. The narratives revealed the origins and meaning of taboos (e.g. food taboos) and rituals and set out norms and rules for the people to follow (e.g. a series of prohibitions placed on menstruating women). They also contained knowledge of the *masalai* and ways in which their powers could be elicited and used by the *kastom lain*. The Paiyamo could call upon the spirits' help in domestic activities such as house building (e.g. the *masalai* could make heavy logs seem 'light as a feather'), making gardens, growing food, hunting or fishing. The *masalai* could also provide them with many magical powers in times of war. They could make distances shrink, enabling people to travel long distances in short periods of time, make warriors invisible and add traction to their arrows. Ayapanai himself is said to have been an owner of an arrow with magical qualities. He could hide behind it and become invisible. Even decades after the Tau house was burnt down, difficult to explain events were often linked to the doings of the *masalai*, and the sites where their presence was known were clearly identifiable in the local sacred geography. The marks left behind as they passed through the landscape were often used by members of the community to legitimise the events described in the narratives and to assert the claim that the stories were, in fact, true.

The first time I heard pastor Jacob speak about what he called 'the secrets of *haus tambaran*' (*amuakemauge* in Paiyamo Amoke), also referred to here as sacred or hidden knowledge, we were sitting in Dinah's house in SBU Camp. It was a Saturday evening, and the pastor was preparing for a community meeting he was hoping to hold after the church service the following day. He brought with him a draft of a document he received from 'a Paiyamo man in Brisbane'. The pastor was clearly intrigued and fascinated by the content of the sheets of paper he took out of his bag, which contained what looked like family trees of the different Paiyamo clans. Each clan, he explained, pointing to the papers, held rights over a clearly defined area of land. The land boundaries were marked a long time ago by the clans' first ancestors and were described in a series of songs and powerful stories (see chapter 4). These stories used to be forbidden and could only be told in the *haus tambaran*. They were very powerful, and if women or children heard these stories, the stories would poison them and they would get sick. In the years that have passed since the burning of the cult house on Tau, however, they have gradually been revealed, eventually becoming public narratives. 'But the true names of the *masalai* from the stories remain hidden,' pastor said, lowering his voice. The Paiyamo told me on a number of occasions that the men used to utter the normally hidden names of the *masalai* before big battles, in healing or when planting new taro or banana gardens. 'Today, the role of the secrets is to serve as evidence of the ownership of land,' I was told by one of the Melefale men as we were sitting in a large, dugout canoe travelling up the Frieda River. The names, I heard my informants say, have to remain hidden until the final land mediation, when they will be revealed. 'At the land declaration, the Paiyamo, Telefomin and Miyanmin will all stand up and call the true names of the *masalai* for the government and the company.

After that, we will not need the secrets. The declaration will be the end of the secret.'

One of the ancestral narratives most commonly told in the SBU Camp was the story of a Melefale ancestor called Melesinapi.[8] Every time the story came up, I was reminded that it used to belong to the *haus tambaran* but was now public. 'But Melesinapi is not his real name. The real name is sacred.' Here is a version of the narrative as it was told to me in 2011:

> This is a story of Melesinapi. But really, it is a story of a man called Wobiru, who lived on the banks of the Awageme River, now called Amosai,[9] close to the current Telefomin village of Wabia. It is a place where many people go to look for gold. The man, Wobiru, used to travel regularly from Amosai, down the Frieda River to Ok Isai and onwards to Seligeme. He was looking for game animals in the bush along the river. One day, the man took his son, Sibago, with him, and together they headed to the junction where Melegeme [Nena River] meets Neigeme [Frieda River]. At the junction, they killed a wild pig. On the high bank of the river, in a place called Kemiti, the father and son built a *mumu* oven from sticks and stones.[10] Once the oven was ready, Wobiru left his son to watch over the *mumu*. He wanted to go down to the river to wash and to cool down. When the man entered the waters of Melegeme, he turned into a crocodile. Starting from his legs and arms, crocodile skin covered his entire body. When Sibago came looking for his father, he saw the crocodile and the crocodile told him: 'Give me one leg of the pig. Then take the rest and carry it to your mother and tell her what has happened.' Sibago left his father and walked away. He started the Yuupaiyamo population in Duranmin, at Awageme. Melesinapi stayed on the junction and is still here. There is a big flat stone at the junction on which the crocodile can sometimes be seen. He gave birth to one egg, which broke and Obilibo was born – the first ancestor of the Melefale. Melesinapi is not his real name. The true name is known only to the Melefale men; it is sacred and kept hidden. Those who know the taboo name can travel to the Nena junction and call the name of the *masalai* to ask him for advice and assistance. Sometimes, if he is happy, he will help them. At other times, he can trick or harm people. He is very powerful.

I heard of people travelling to the Nena junction to talk to Melesinapi. In order to do so, they had to call his secret name. Only then would the *masalai* come down and listen to them. Three elements were particularly significant in the way in which the Paiyamo elicited and talked about their encounters with the *masalai*. Firstly, by travelling to a sacred site and uttering a sacred name they enacted a belief in the *masalai* and the teachings of the *haus tambaran*. 'When people talk to the *masalai*, they need to do so with belief. If they do not believe in the spirits, the spirits will have no power and they will not be able to help them,' it was explained to me by one of the women in Paupe. A number of times I have heard a version of this statement linking the *masalai* and their power directly to the sincerity of people's beliefs and acts of worship, where 'belief' describes a form of commitment with an implied future orientation. Secondly, the calling of the name made the relationship between the man and the *masalai* visible. By demonstrating his knowledge of the name, the caller revealed himself as a member of a *kastom lain*, thus validating the grounds for the relationship.[11] In this context, knowledge served to mark a boundary between different groups and lineages among the Paiyamo, placing the knower of a name within a particular group – with all the rights and responsibilities this membership entailed. Thirdly, my informants in Paupe could never be sure what the *masalai* would do. The spirits were at once feared and trusted, mischievous and respected, powerful yet unpredictable. People tended to give offerings to the spirits and generally tried not to upset them with their words or actions – but to a large extent, the outcome of their relations with the *masalai* depended on the good (or not so good) will of the latter. Although sites of the spirits' presence were clearly marked in myths, stories and landscapes, the *masalai* would not always be there. Sometimes, I was told, they would go deep inside the bush, and people did not know exactly where they were or what they were doing when they were away from their usual places.

Not long after I heard the story of Melesinapi for the first time, I sat down to talk to another prominent member of the Melefale. Peter was Tikap's son and, alongside pastor Jacob, was actively involved in engaging with the company in the process of negotiating agreements with regards to the Frieda River Project. We talked about the different groups that make up the Paiyamo Sane and about the stories they tell of their origin. It was a hot and sunny day, and Peter was relaxing in the morning sun, smoking a cigarette and sharpening the blade of his bush knife with a stone he had found in the river. Before long, the conversation turned to Melesinapi. 'There is this crocodile,' Peter told me. 'He stays at the Nena-Frieda junction. He is the first ancestor and the *masalai* of the Melefale. People call him by his public name, Melesinapi, but his secret name is *XXXX*.'[12]

Let Me Tell You a Secret

In the introduction to this book, I ethnographically placed the Paiyamo on the fringes of the Mountain Ok area, also known as the Min region of Papua New Guinea. In the last quarter of the twentieth century, numerous anthropological studies from this part of the country described intricate local systems of distribution and management of knowledge based on patterns of concealment and revelation.[13] A number of ethnographic accounts from the region focused on initiation cults performed by men in which secrecy and deception were pervasive. Here, possibly the best known example is

Fredrick Barth's widely read study of *The Ritual and Knowledge among the Baktaman of New Guinea* (1975) in which he described a welfare-preserving cult performed by men and dependent on effective management of 'secret' or 'sacred' [*awem*] knowledge. At the heart of the Baktaman religious practice was a sequence of seven degrees of initiation separated by indefinite periods of time. At each stage, elements of sacred knowledge were revealed to novices, placing them in a ritualistic cycle of carefully managed concealment and strategically timed revelation.[14] As described by Barth, the value of *awem* knowledge depended on it being revealed to novices during sacred initiations as much as it depended on it remaining hidden.

This interplay between what is hidden and what is revealed to whom and under what circumstances characterised traditional religious practices of the Min, where 'sacred secrets [were] meant to be told' and where 'telling secrets to novices [was] one of the main tasks of the men's cult' (Jorgensen,1990a: 41). But anthropological accounts from the region suggest that the ritual moments of revelation tended to elicit in novices feelings of puzzlement and confusion rather than increased clarity. Barth (1990: 642–3) wrote that the Baktaman initiation rituals included 'staged-managed revelations' where cult members made secrets and mysteries 'immanent – but not easily understandable – to a congregation or a handful of overwhelmed novices'. Manipulating symbols to construct 'a complex and moving tradition of knowledge', they then used it to 'affect the novices by its force, not simply to explicate the knowledge to them'. Reflecting on his work among the Miyanmin, Don Gardner (1983: 352) described how all Miyanmin initiation rituals

> involve the revelation of ancestral relics . . . the utterance of secret, sacred names for plants, animals and cultural artefacts given by the original ancestors, and the singing of sacred songs. Many rituals contain mythological accounts of the origins of the [Min] and their customs. This lore, however, is never presented in its entirety and it tends to bewilder the initiands rather than enlighten them.

Dan Jorgensen used the term 'turning' to explain how in each subsequent initiatory rite among the Telefomin 'new esoteric renderings shift the sense of previous secrets' (1990a: 43) and 'the obvious becomes trivial while the improbable acquired significance' (1981: 434). He articulated a paradox at the centre of this process, which is 'that men, in their roles as initiators, say that they tell secrets in order to make novices see; but when speaking about their experiences as novices, they say that the initiators "turned things" and kept "turning" them until one got dizzy' (1990a: 41).[15] Such 'turnings', Jorgensen claimed, entailed adapting to new frames of reference, which enabled for conventional meanings to be displaced and substituted with new ones and for the sacred knowledge to be revealed. Writing about his work among the Angkaiyakmin in Bolivip, Tony Crook (1999: 239) used a similar expression, as he recorded initiates saying that 'when senior men "turn their words around" the juniors feel confused, often saying that their heads just spin with all these different possible paths of talk moving around'.[16]

There are many similarities between the ethnographic accounts from the Min region and the way people in Paupe talked about their traditional religious activities. Like many of their Mountain Ok neighbours, they too practised what Barth (1975) described as a welfare-preserving cult performed by men, which pivoted around taro gardening and warfare and where access to religious knowledge was carefully managed and granted only to the *kastom lain* through sporadic ritualistic revelations. As in the cult houses of the Min, in the dimly lit spaces of the Paiyamo *haus tambaran*, the telling of secrets was often accompanied by the presentation of sacred objects. For example, the telling of the story of Semiti was accompanied by a presentation of a piece of the stone from Mt Usimi. Such presentations served to evidence the story, complementing what Jorgensen (1981) described in the Telefomin context as 'knowing through hearing' with 'knowing through seeing'. By the time I arrived at Frieda, there were only two men still alive in Paupe who claimed to have entered the house on Tau Mountain. Although their memories of it were becoming blurry, they could still remember the sense of excitement and mysticism and the presence of great power that accompanied much of the cult's religious activity, as is prominent in accounts from the Min region. They could also recall the sense of bewilderment they experienced when the elder men taught them the sacred knowledge of *haus tambaran* – but this was not the purposeful 'turning' of the kind described by Barth, Jorgensen, Crook and Gardner; rather, they were stunned when things that seemed very familiar were revealed to them as having been something else entirely all along.

To an extent, this difference might be explained by the fact that the Paiyamo cult was not centred on a sequence of distinctive initiations, each uncovering the next layer of esoteric knowledge. When asked, the Paiyamo would say without hesitation that the tradition built around numerous initiatory stages, deception and 'turning talk' belonged to the Telefomin and the Miyanmin but that they themselves were not a part of it. The Paiyamo performed only a single prescribed initiation ceremony, which took place when new novices entered *haus tambaran* for the first time. It involved hitting novices' skin with stinging nettles and rubbing it with the fat of a wild pig mixed with some sort of red ochre and planting a new taro garden. Tikap once told me that although this formal initiation ceremony was important and constituted a rare and significant event in the life of the community, young men were in fact initiated every time a situation arose that required the cult's intervention, and with which they were not familiar, and whenever *kastom lain* had to teach its newest members something new in response to a given circumstance. In those activities, they did not use purposeful deception. My informants in Paupe frequently expressed their frustration with the way they felt that Wabia and Ok Isai communities continued to 'turn their talk' in the context of the Frieda River Project. 'The Telefols have always been known for turning their talk and changing their stories,' they would say. 'In the past, this process belonged to the Telefomin cult house and to the cult of Afek. But now the Telefols are serving a very different cult – one of development and money. But not us. We, the Paiyamo, we don't turn our talk. We still have our secrets, but we speak directly.'

Listening to the stories the Paiyamo told about the sacred knowledge of *haus tambaran* and the names of the *masalai*, which needed to remain hidden, it becomes clear that treating Paiyamo 'secrecy' just as a distributive system of knowledge grounded in a religious cult performed by men gets us, at best, part way towards seeing what these secrets were and are about. In the teachings of the *haus tambaran*, sacred knowledge was knowledge capable of eliciting a particular effect in the world (aimed at ensuring the survival and welfare of the community), which made it valuable and powerful. But the value and power of the sacred was judged not, to borrow Crook's (2007b) phrase, against the truth of the knowledge but its efficacy. Having spent time among the Paiyamo, I came to understand that the true subject of revelations in their *haus tambaran* was not the content of the sacred knowledge but its meaning and the way in which its efficacy could be elicited and managed. In this interpretation, what differentiated the *kastom lain* from the rest of the society was not so much that they knew what the sacred knowledge said but that they knew what it did and were able to utilise its power. To some extent, at least, this may account for reports of women, children and uninitiated men knowing the sacred names and stories. It can also explain why, decades after the last initiates entered the Paiyamo *haus tambaran* for the first time, Peter told me the hidden name of Melesinapi seemingly carelessly very early on (but was reluctant to do so again only a few months later once I developed a more in-depth knowledge of local history and *kastom*) and why the Paiyamo saw no real need to keep such information hidden from the company developing the Project. As a prominent Melefale man explained to me, the utterance of the sacred on its own had no power:

> At times, people know the names and stories and do not even know that they are secret. Because the company does not understand the value of the sacred knowledge, they cannot affect it; it is useless to them. If they are willing to know, we can tell them the hidden name of the *masalai* and this would not diminish the spirit's powers . . . but the name has to be kept hidden from the other Paiyamo clans because if they knew it they could use it against us or take away the power of the spirit.

This last statement is reminiscent of comments made by Roger Lohmann (2000, 2001 and 2003) in his description of religious practices among the Paiyamo's closest ethnographic neighbours, the Asabano of Duranmin. As mentioned in the introduction, the two groups tell a similar story of the first ancestress and her mother, who went on to Telefomin and became known across the Min region as the great ancestress Afek. The two languages are closely related, with Robert Conrad and Wayne Dye's 1975 study finding the 'Paupe language [*sic*]' to be 29 per cent cognate with the language spoken in Duranmin. Members of both groups describe close historical ties between them, with a high number of individuals among the Asabano explicitly identified by my interlocutors in Paupe as the Imlou, the sixth clan of the Paiyamo. Although at the heart of Asabano traditional religion was an initiatory cult framed around eight distinctive stages of initiation, closely resembling practices of the Min rather than the Paiyamo, both groups spoke of shared elements in their traditional religions and claimed that, in the time of ancestors and *kastom*, they were able to participate in each other's rituals and enter each other's *haus tambaran*, with the Melefale mythology speaking of a cult house at Tau Mountain that they shared with the Imlou (see Lohmann 2000 for the ethnographic material on the Asabano). Like in many areas across the region, Lohmann (2001: 96–97) argued, the Asabano religious knowledge had to remain hidden; this secrecy characterising traditional religious practice and rendering it 'powerful, psychologically and magically, by restricting its circumstance of enactment'. He further described how 'the Asabano protected the secrecy of their religious knowledge by proclaiming it not only useless to the uninitiated, but also subject to complete power loss or disaster should secrets be inappropriately spoken or revealed'.

The Paiyamo used a version of this proclamation in a literal sense – to them, hidden knowledge was *actually* useless to the people who did not have the ability to understand its true value and who lacked the relations and endorsements through which they could affect and use its power. In the past, in the time of *haus tambaran*, people who lacked the ability to grasp the meaning of the secret were indeed seen as 'the uninitiated'. But with the initiations long left behind and present only in ancestral stories and fleeting memories of the oldest men in Paupe, it was not a sole prescribed initiatory ritual that separated those who did know the secrets of the Paiyamo from those who did not, if indeed it ever was. It was the quality of relationships between people – judged against the ways those relations were made visible through material and immaterial exchanges embedded in care obligations and based on respect and trust – that spoke to dividual aspects of Paiyamo personhoods. The true meaning and efficacy of knowledge became known to people over time, through experiences they shared with others and the quality of social relations they formed with them. In some contexts, those relations pivoted around the anticipation of an operating mine at Frieda River, emphasising the role of secrets in marking group membership and evidencing clan ownership of land. The Paiyamo often talked about the role secrets will play at what they referred to as the land declaration. During that event, they told me, all impact communities will stand up and call out secret names of the *masalai* in the project area. Whoever calls the right name will be revealed as the primary landowner in the Frieda River Project area. 'The company already knows what this name is. They have a book and it is all written there,' which I was told on at least four separate occasions by people in Paupe as well as Ok Isai. At the same time, my informants would often remark that the company did not care to know the secrets (meant in this context as the hidden names of the local spirits), instead focusing on mapping out land boundaries and collecting genealogies and census data in the impact communities. 'I do not think they believe in the *masalai*,' I was told by one of the older women in Paupe (see also chapter 4).

The issue of belief holds the key to understanding the Paiyamo practices related to the management of sacred, religious knowledge. The belief in the efficacy of the sacred knowledge is what gave it its power, even in what my hosts referred to as 'the time after *kastom*'. I heard in Paupe that 'if the hidden name is revealed to other groups, if we stop keeping secrets and believing in spirits, the *masalai* will lose their powers and eventually become nothing more than another creature in the bush'. The practice of secrecy, of keeping the names of the *masalai* hidden, constituted an enactment of elements of a traditional religious doctrine, of a belief in the powers of the *masalai* and the ways those powers could be elicited and directed. For example, by then the story of Melesinapi constituted a public and popular narrative among the Melefale, and even the children could be heard playfully telling each other make-believe stories of the crocodile from the Nena junction. But the Melefale would often say that maintaining the secrecy of his real name was what made the spirit powerful; if this practice was abandoned and the *masalai*'s name made public, his powers would diminish and in time disappear. As such, the phrase 'this is secret', whispered in a lowered voice, was a common feature of many coffee nights spent around warm hearths in Paupe.

The Holy Spirit Came and the *Kastom* Had Finished

To introduce the topic of major religious changes experienced by the Paiyamo in the 1970s, let me return briefly to the story of Semiti and to Mount Usimi, where the great ancestress gave birth to a powerful stone. As I described in chapter 2, in the past any man who knew the hidden name of the *masalai* could go to the mountain and pick up a piece of that stone. The pieces were small, fitted easily in the man's *bilum* and could be carried effortlessly off the mountain by just one member of the *kastom lain*. A piece of the stone would be shown to novices when they entered the *haus tambaran* so they knew that the story was true. This stone had many magical powers, but most important was its ability to deliver a good harvest, thus helping to provide food and ensure the welfare of the community. Men used to plant pieces of the stone under a red tanget plant whilst chanting the name of the *masalai*. With the spirit's help, the garden produced an abundance of food and was protected from misfortunes such as landslides or attacks by wild pigs. Some people in Paupe say that 'now that we have heard the word of God and Jesus, the stone has become too heavy for one person to go, take a piece, and carry it down the mountain'. Many others say that this stone birthed by the ancestresses has since turned to gold but maintains its great powers. An increasingly popular view among the Paiyamo is that it was God who buried the gold in this geographically remote area: 'People believe God gave them minerals so that they can use them. Because God put us in mountainous areas, he could not give us plenty of fish and other resources, so he gave us gold and copper instead.'

According to local narratives, the Paiyamo first learned about Christianity and God in the early 1970s when an American Assemblies of God missionary from the Sepik arrived in Paupe. Although some members of the *kastom lain*, including Aiyapanai, refused to listen to his teachings and covered their ears so as not to hear the Christian songs and prayers, most of the people welcomed the new arrival. They were not violent and did not oppose

Illustration 3.2. Front entrance to the Baptist church in the SBU Camp. Photograph by the author.

Illustration 3.3. *Bilums* hanging on the wall in the church at the SBU Camp. Photograph by the author.

the missionary. He brought with him 'the word of God' and they were curious. But it was not until a few years later that, the Paiyamo say, they really started to 'hear' this word. In the late 1970s, a well-documented religious revival movement swept across the Min area. It was led and orchestrated by pastor Diyos and 'acted to entrench Christianity within society at the same time as it precipitated the fervent destruction of indigenous ritual paraphernalia, procedures and structures' (MacDonald 2013: 12).[17] Roger Lohmann met Diyos and was able to capture the story of this religious revival as told by the pastor himself. It is an interesting and deeply personal description of a man's conversion, vocation and missionary work, characterised by acting on messages received directly from God (Lohmann 2000, 2007). In 1974, Diyos set up a bible school in Duranmin that came to be called the Sepik Baptist College, and it is there that in 1977 a charismatic outbreak took place, triggering a religious revival that subsequently spread around the region. Diyos described that moment to Lohmann (2000: 198–99) in the following way:

> When I began the prayer, the fire of God came through and hit me and went through the people. It came through like lightening. And they saw the fire come straight to my shoulders and through like that. And all the men and women fell down. As they lay there they cried, and my wife and those who were angry were rejoicing: the big fire had risen! It was something else! We were happy and singing songs, and many men

> and women were lying about the floor, and some had collapsed outside the church. They confessed their sins and sat there crying. I kept walking and didn't sleep. I went to the radio and told the missionaries what had happened, what father God had done this night. Lots of missionaries came: three planes came to Duranmin the first day. They came to see this, all these people lying where they had fallen. They were falling down and crying, and bringing all sorts of their religious relics; human bones and other things . . . At that time, Satan's power was broken. The power of Afek was broken at that time. Every day after this, women could go inside the spirit houses. They went and looked inside and saw what things were there. Children would go inside. Everything that had been taboo for women to eat they were given to eat. Now everything became free because God supressed the power of Afek at this time . . . Before this time, according to the old beliefs about what Afek declared, it was a big deal, the stories about these taboo foods and rituals. Should all the food be finished, with no power, everyone would die. Because people believed this, when I was preaching, I had to look out for the beliefs of the important men. If a grandfather was there, I had to avoid him. Now I don't have to avoid them, I can go straight and talk clearly to the people, and they can understand quickly. Now the belief in God has taken over.

Among the people who witnessed those events in Duranmin was pastor Jacob, who started his training at the college that year, having converted to Christianity in an Assemblies of God-run school in Ambunti a couple of years earlier. He remembers seeing people shake and collapse and hearing them speak in tongues as God entered their minds and bodies. He also remembers hearing about the *haus tambaran* on Tau being burned down and the missionaries proclaiming the arrival of the Holy Spirit and the victory of good over evil. Shortly after the events in Duranmin, he decided to travel to Paupe to bring the Holy Spirit to his family and the Paiyamo. One sunny morning, as we sat outside Dinah's house following an unusually short Sunday service, Tikap, with his white beard and fading blue eyes, said to me:

> When we heard the word of God, we were all rejoicing, and we were quick to abandon the *kastom* and join the church. Joining the church meant that we could all eat taboo food. Jacob and the men from *haus tambaran* cooked all kinds of food for everyone. We all prayed to God. Then the *kastom lain* gave the previously tabooed food to women and children to eat, and no one got sick; everyone was fine. And then we knew that the Holy Spirit came and that the *kastom* had finished.

Jacob finished his training at the bible college in 1979. In 1980, he returned to Paupe to assume his pastoral duties and has since become an established religious leader of the Paiyamo community. He married Lina, a Telefomin woman from Ok Isai, in 1982. Dinah was the oldest of their four children.

People in Paupe were happy to discuss with me the importance of the Bible. They described themselves as 'the children of God' and were likely to burst out in spontaneous prayer at all times of day and night. The attendance at Sunday church services was generally high, and people's participation in Christian rituals was eager. On a typical Sunday, one of the church leaders would begin the service by reading from the Bible and delivering a sermon. A number of community members would then get involved in the conduct of the service, reading Bible passages, giving personal testimonies of faith, expressing gratitude and asking for God's help and assistance. Pastor Jacob told me on numerous occasions that although the Paiyamo used to draw their powers from *kastom* and from the *haus tambaran*, they were now a Christian community. 'Now we believe in God and the *kastom* is no more,' he would say repeatedly. But on other occasions, he would remark that as the son of Aiyapanai he knew the stories and secrets of the *kastom lain*. He was also the first person in Paupe to teach me about the continued importance of keeping the names of the *masalai* hidden. He was never initiated into the cult, instead pursuing Christian religious education. However, his father taught him many magical spells and passed on to him much of the powerful sacred knowledge. As a pastor and a Paiyamo leader, Jacob said that he now passes on to the next generations the good *kastom* only; that which agrees with the Bible and the word of God. He made it clear that there is also bad *kastom*, powerful, destructive spells that are not passed on in the hope that if they are not practised they will be forgotten and their damaging powers will be lost.

In the absence of an organised cult, it was up to the fathers such as Jacob to pass on the traditional sacred knowledge to their sons, but many were reluctant to do so. 'Because most people now go to church, the elders don't like talking about the past,' a young Paiyamo man once told me, expressing a sentiment I came across many times in Paupe – one where young men longed for knowledge of magical spells and powers of the cult house but felt that they were denied access to this knowledge by their now Christian fathers and grandfathers. At the same time, most men I met in Paupe talked about how they came to know the sacred stories and secrets of the cult house. Peter told me:

> My father would sometimes call me over and say: you come to me after I come back from the garden. There is a story I want to tell you. He was worried that if he doesn't tell me the story will die with him. Every man is trying to find someone he can trust to pass the stories and secrets on to. And I would go to my father, hear the story and learn.

'I am Christian and I worship God. With spirits, I just know the names and stories,' he added. I pulled him up on this statement some time later, after I heard him tell another Melefale man of his plans to go and see Melesinapi. Peter had not been feeling well and was hoping that the *masalai* would tell him why he got sick. I asked him whether he thought that God would be angry with him for seeking out and speaking to the *masalai*. He smiled at me and said: 'But why? When God created earth, from day one he created the spirits, too.'

There are a number of anthropological accounts from the region that look at the ways the communities in the area made sense of Christianity and placed it in relation to their prior religious beliefs. Joel Robbins (2007) described how the conversion constituted a rupture in the world of the Urapmin, who, following a direct experience of Diyos's religious revival, claimed that when they became Christian everything changed. In this context, Robbins (2001) saw Christianity not as an independent force but rather as a set of institutionalised promises. These were welcomed by the Urapmin, who consequently rejected their *kastom* with vehemence to embrace this new, modern potential. The Urapmin, as Robbins described them, were never interested in syncretising Christian and traditional ideas. After the religious revival, they embarked on a project to 'rework their lives in what they understood to be solely Christian terms . . . informed by modern promises of personal and social transformation' (2001: 903). Although occasionally they reverted to *kastom* magic and rituals, they interpret these instances in terms of backsliding and sin and not as 'moves between religious and political factions' (1995: 214, see also Robbins 1998a and 1998b). Roger Lohmann focused on the concept of cultural reception and the role it played in the Asabano conversion to Christianity. He argued that despite the destruction of sacred objects, desecration of religious spaces and explicit proclamations of the victory of God over the power of Afek that accompanied the religious revival, 'in general, the traditional cosmology has not been abandoned and replaced with Christian cosmology. Rather, it has been couched within what Asabano understand to be the broader, deeper, universal truth of Christianity' (2000: 183). Lorenzo Brutti (2000: 109) described conversion to Christianity among the Oksapmin as one element of a broader, complex and ongoing process of innovation, described not in terms of a cosmological break with their traditional beliefs but as finding a way to incorporate dramatic transformations in the frame of their original cosmology, 'sometimes by restoring it, always through a plastic operation of redefinition of cultural models of cosmological explanations'.[18] For the Oksapmin, he argued, Christianity did not mark a point of rupture or a break from their traditional beliefs. Rather, it served as a catalyst that altered the existing religious processes and structures to accommodate the new potential introduced to the people with the promise of modernity and material gain (Brutti 2007). A few years later, Fraser MacDonald (2014: 180) offered his own interpretation of this process by suggesting that 'throughout history, the Oksapmin have adjusted their cosmology, sometimes profoundly, in order to marshal new flows of sacred power and wealth, thus rendering their religious world an eclectic, composite structure made up of elements both indigenously created and exogenously borrowed'. Moving away from Robbin's theory of religious duality and Brutti's notion of religious superimposition, he suggested an interpretative model that he branded 'religious fusion' and defined as 'a creative process of mutually adapting and transforming the Christian and indigenous worlds in the formation of a single, composite religion' (2013: 17) – embedded in 'a dual process of mutual adjustment – the indigenisation of Christianity and the Christianisation of indigeneity' (ibid.: 109).

The Paiyamo clearly differentiated between sets of religious elements they associated with traditional beliefs and those they associated with Christianity (or *lotu* as they referred to it in Tok Pisin). Pretty much all the people I spoke to in Paupe about their identity as 'the children of God' talked about how they abandoned *kastom* when they received the Holy Spirit. They would describe *kastom* as 'that which belongs to the ancestors' and Christianity as 'that which belongs to the present'. They could also be heard proclaiming that they were no longer afraid of the spirits because now God was looking after them. At the same time, they would openly remark that sacred ancestral stories were still important and that the spirits continued to be an inherent part of their world. They told traditional stories alongside biblical narratives and turned to the practices of their ancestors as well as Christian rituals and prayer when dealing with key moments in the life of the community, such as death of one of its members. A Melefale elder, Pawi, explained to me at one point that 'we may not like it, see the *masalai* as Satan and his power as sin, but they are all around and are not going anywhere', with pastor John, leader of the Inaku group, adding:

> When the missionaries came we found out the talk of God. People discovered that food taboos were a lie and they were happy. But the spirits are still here. They are not something that will end. If we do not worship them, though, they will have no power. They will just stay in the bush and look at us from outside. But they will still be there.[19]

The Paiyamo seemed able to reconcile the two religious registers – traditional beliefs and Christianity – with relative ease, which is not something I could say about my own experiences of trying to understand their contemporary cosmology and what can be referred to here broadly as their religious lives. Statements that I found puzzling and sometimes contradictory did not have the same effect on my interlocutors, who moved freely between stories belonging to *haus tambaran* and Christian narratives when trying to interpret and at times direct and influence events and the relations affected by them. In this context, there were three ethnographic moments that intrigued me the most – and I turn to them here. First is a story I heard at the SBU Camp about Melesinapi disrupting the work of the company developing the Project. Second is a conversation I had with Rhonda in which she reflected on herself as a Christian. Third is a set of observations and conversations I had about death and an afterlife in the difficult context of witnessing the passing of a young boy in Paupe. While the first two take the form of short snapshots of my field material, the third one is longer and contains within it several different narratives of death and an afterlife.

Moment 1: The Masalai and the Company

At one point, during a conversation about a particularly mischievous *masalai*, I asked my hosts in Paupe whether the spirits were happy for the company to conduct exploration work in the area. In response, they offered me this story:

> In 2009, the company constructed a camp near where Melesinapi sleeps on a stone at the junction between Frieda and Nena rivers and called it the Frieda Bend Camp. Although the company employed some local people to work at the camp, they experienced many issues related to the safety and wellbeing of the personnel (with multiple instances of sickness and injuries) but also equipment malfunctions, which halted the plan of work and required expensive repair. This is what happens if the *masalai* is not happy about being disturbed by the company. He makes the workmen sick or breaks tools and machinery. He can send bad weather so that work cannot be done. For example, when the company continued to repeatedly hire the same man to transport people and goods in a motor canoe, that man got sick. On another occasion, the company was drilling very close to the junction and their drill broke. Another time, not that long ago, the spirit got angry with some company employees and many men fell ill with fever and diarrhoea. When Melesinapi made the workmen ill, we felt sorry for the company and for the workers. We went to the junction to plead with Melesinapi to let the company work – we were a group of men and women; some children, too. When we got there, some of the men called the *masalai* by his hidden name and offered him food. Most times, we kill and offer a pig, taro and other garden foods to the spirits. But that time the company provided chicken and rice, which was offered to Melesinapi and eaten by the people. The pastor led a prayer asking God to help the workers and the company. This is what we did. We called the spirit by his secret name, offered him food and prayed to God asking for his support.

This is where the story ended. Whether or not the problems in the camp stopped as a result of this ritualistic performance seemed less important to my companions than the performance itself. Intrigued by the interplay between what could be seen here as distinctive traditional and Christian religious elements, I asked the storytellers which was more important in this particular situation – the power of the spirit or the power of God. Somewhat surprised and confused by my question, they explained to me that for them to be able to ask Melesinapi's help, *kastom* and Christianity had to work together.

Moment 2: Oneibo

Rhonda's house was located on top of a small hill at the SBU Camp, only a couple of minutes' walk from Dinah's household. The house was raised about a meter above the ground and situated at the end of a steep path leading up directly from where I stayed with Dinah and Mawi. It was made entirely out of bush materials, although large cardboard boxes carried over from the Airstrip Camp were used to line its walls to provide insulation during the day and serve as sleeping mats for the numerous inhabitants at night. In the middle of this simple, wooden structure with its woven leaf roof and a palm bark floor was a clay hearth where food was prepared and where water was boiled, sometimes numerous times a day, to make very weak and incredibly sweet tea. I spent a lot of time in that house learning to make *bilums* and partaking in the household's daily activities.

Rhonda was the daughter of Tikap and pastor Jacob's sister, Ciare. She was very outspoken about her faith in God and was considered one of the key female leaders of Paupe's Baptist community. She was a regular Bible reader, and when she was not at work at the company camp where she was employed as a laundry assistant, she presided over women's services held in the SBU Camp's green church building on Wednesday mornings. She described herself as a Christian woman and prayed regularly, directing all her worries and all her joys to God. 'These days we listen to the word of the big man (here used to mean God). We see the *masalai* as bad spirits. They are a belief from the time of our ancestors, not ours,' Rhonda remarked, looking up from an intricate pattern of one of the many beautiful *bilums* she worked on simultaneously at any given time. In these statements, she clearly separated *kastom* from Christianity, aligning herself with the latter. But on one particular morning, as we sat on the well-worn floor of her kitchen preparing a simple taro meal for her grandchildren, she looked at me and said: 'I am Christian and I believe in God. But I have heard the bird twice.' With that, Rhonda began telling me a story of a little bird called Oneibo:

> He is a bird and he is also a *masalai*. He dwells deep inside the bush but can sometimes come close to the village. For example, when you are away and staying in another

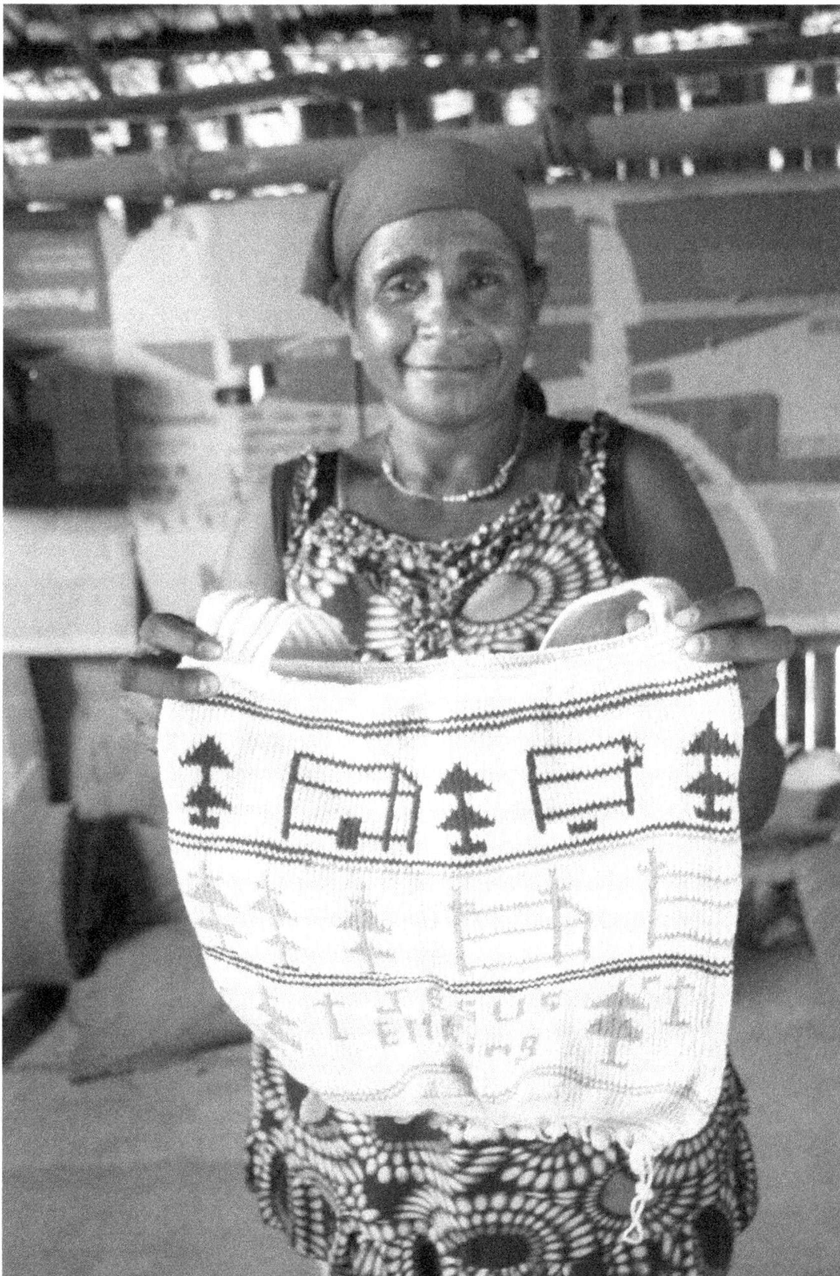

Illustration 3.4. Rhonda holding one of her famous *bilums*. The inscription reads: 'Jesus is King'. Photograph by the author.

> community, he can visit you there and bring you messages. When he calls you, you will not be able to see him – he is small, has a black belly and a long nose. But you will hear his song, a message, and you will understand. The first time I met Oneibo was shortly after my mother's death. I was in the bush with my husband, who has since died. We left our young daughter in the village in the care of another woman. 'Hurry up and go back to the village' – the bird told me – 'your child is sick and is crying.' He repeated this message three times. I tried to ignore it, but finally I gave in and decided to listen to the bird's words. I told my husband we should hurry up and go home. We returned to the village, and we found our youngest daughter sick and crying. I now know that Oneibo's voice was a message from my dead mother, looking out for me and the baby. Another time, I was in the bush with some other women. We were washing sago[20] when we heard the bird calling: 'You are in the bush washing sago and I have come to the bush, too.' On our way back to the village, we were met by messengers from the community carrying the sad news that one of the local women passed away that day. So I knew it was her who came to see us in the bush. Like all *masalai*, Oneibo can carry true messages. When needed, he can come to help, and even to the rescue. He is often heard in the bush calling out a warning, asking people to follow him or to return to the village, but he can also play tricks on people; when you hear his song it is sometimes hard to know which one is which.

Although Rhonda frequently said that she now believed in God and that her Christian faith opposed the traditional religious practices of worshipping the *masalai*, she has heard Oneibo and experienced his presence. She recognised him from the stories she used to hear and realised that those stories were true. The fact that on both occasions the bird's messages related to events that had taken place some distance away offered further evidence of Oneibo's powers. She did not see it as a bad thing or a thing that she could control. Nor did she apologise for following the *masalai*'s word.

Moment 3: The Death of Navi

It could be argued that few events in the life of a community elucidate different aspects of their beliefs and religious practices as clearly and vividly as the death of one of its members. In all religions and cultures, death, with all its ambiguities, marks a rite of passage from one state of being to another and forces to the surface aspects of cosmology that may often lay concealed under an array of daily trivialities. In the past, before Christianity introduced the concept of heaven, a raised wooden structure was built in the bush following the death of a Paiyamo person. The body of the deceased was laid out on top of the structure and left there until it decomposed. If the dead person was a prominent figure in the community, their bones and a skull would then be collected, put in a *bilum* and kept in the *haus tambaran*, while the dead themselves would join the *masalai* in the bush.[21] If a living person wanted to speak with them, they had to go into the dense rainforest, call the *masalai*'s hidden name, call the name of the deceased and ask if that person was with him. If the *masalai* wanted to help, he could make communication between the deceased and the living possible. While the villages were places of the living, the bush and the river were often perceived as places where the dead and the *masalai* reigned. It was common for the dead to wander around the bush, and they have been known to enter the villages at night-time, walking between houses and listening in on people's conversations. But they would never enter the houses. Inside the houses, I was told, people were safe.

Since then, Christian doctrine has infiltrated the local narrative of an afterlife. The Paiyamo told me that because they have now heard the word of God their dead no longer went into the bush. Instead, if they were a good person, they went to heaven; if they were bad, they went to hell. But although the wandering dead were openly declared a thing of the past, stories of travelling ancestral spirits were still being used to explain some of the otherwise hard to comprehend happenings in the village. To give a brief example, one morning I woke up to a commotion outside the house. A number of women were engaged in a very animated conversation. The previous night, a young boy living in SBU Camp, woke up needing to pee. He roused his mum and they walked over to the door, careful not to wake any of the other people asleep in the house. The woman opened the door and stood there watching while the boy went behind the house. Suddenly, he saw a man. The man was old and thin, and his hand was outstretched, reaching for the boy. The boy screamed. His mother told us that she too opened her mouth but the spirit closed her airways and only a quiet, muted sound came out. Her son's screaming woke Peter, who was sleeping in a house nearby, and he came out to see what was happening, but by that point the boy and his mum were already heading back inside the house, where they knew they would be safe. The next day, the SBU Camp was full of talk. The conclusion reached by an early evening fire as the day was drawing to a close was that the man was the boy's deceased ancestor who was buried in the village near a place where the child was seen playing a couple of days earlier.[22] As the incident happened shortly before daybreak, the most likely explanation was that he must have disturbed the old man when the man was coming back from the bush at the end of the night. As was often the case, whilst illuminating aspects of what can be seen as the traditional beliefs still vividly present in the consciousness of my hosts at the SBU Camp, the evening concluded with a proclamation that 'in the past dead people went into the bush, or so we thought. But now we know better. We know that they all go and stay with God in heaven'. Dinah and other people I later spoke to about this event seemed aware of the apparent contradiction between the two narratives but were noticeably amused by my own inability to see its irrelevance to the understanding of the event itself and the conversations that followed.

There was one death in Paupe during my time there. A little boy who I call here Navi was taken ill and brought to the Airstrip Camp – the nearest point of access to services for the Paupe community. Two medical professionals were sent from the Base Camp to help the child and his family. By the time I arrived at the old Community Affairs office, around two dozen community members were already there. Navi was receiving fluids through an IV drip and was being given oxygen to aid his breathing. 'This baby doesn't need oxygen, it needs to be admitted to a hospital,' one of the doctors told me quietly when I inquired about the boy's condition. This statement and the look in the doctor's eyes conveyed a sense of sadness and helplessness that was mirrored in many other faces in the room. Although the medics were happy to help Navi, they talked to me about the difficulty of their position. 'As company medics we are not really at Frieda to help the communities . . . we are here to look after the workers. This is our job and our priority. It is the government, not us, that should be providing health services to these people.' The two doctors were worried about creating a precedent and enhancing the already prominent structure of dependency on the company that had developed in the area (see chapter 5). They were also concerned that such a dependency placed constraints on their ability to help the people they had a direct (contracted) duty of care towards; being with the Paiyamo entailed not being with the company employees. Despite all this, and with management's approval and support, the two doctors spent the whole day at the airstrip's old Community Affairs office, with one of them extending his stay until the next day. By late afternoon, all the necessary arrangements had been made by the company for Navi to be medevacked to Wewak first thing the following morning.

Some of the women that had kept vigil by Navi's side left the office and returned a short time later, bringing with them food and water. At dusk, pastor John led the gathering in a prayer that called for mercy from God and strength for the little boy and his family. People were asked to speak out if they held some heavy feelings towards another. They were told to apologise if they had any apologies to make. 'We need to resolve all our problems and end all our conflicts, and we need to do it now,' said the pastor, with voices around the room echoing this sentiment. After the prayer, people resumed their vigil, which was interrupted only by individual prayers and quiet, short verbal exchanges. One by one, they approached the boy and touched his skin, sharing their strength with Navi in the long wait for the morning flight to the hospital.

Navi did not make it on to the morning flight. He died in the early hours, just as the sun started to rise above the hills surrounding the airstrip. Company workers provided some petrol for Navi to be taken home to Aston in a motorised canoe. Cries, wailing and dramatic chants resounded up and down the river, carrying with them the sorrowful message of the boy's death. A house of mourning was set up in Navi's uncle's house in Aston. The boy's body was brought into the house where it was washed, wrapped in decorative fabric in preparation for burial and laid down on the floor at the centre of the structure. His closest family gathered around him, with other members of the community slowly filling up the remaining space. Dinah explained to me that the family of the dead will stay in the house for a week:

> In the past, it was two weeks but now many people have work obligations so that time had to be shortened . . . in this time they will not cook or enter the garden, and others will look after them for food and drink. They will not wash except for wiping their faces and hands with water brought to them by members of the community, and they will not change their clothes. After a week, the family will wash, get changed and provide food for everyone. We will have a feast and close the house of mourning.

Many of the people living in Paupe spent a significant amount of their time in the house of mourning, telling stories of *kastom* and the ancestors, remembering Navi in narratives and anecdotes and coming together in prayers led by local church leaders. Dramatic cries, screams and laudations that in the past were used to help the dead on their way to the afterlife were now actively discouraged, and I overheard Rhonda telling some women off by saying: 'You cannot cry too much – this will disrupt the road to Heaven for him. We are now Christian; we cannot cry and worry too much. It was God's will and we need to accept it.' But she too joined in, weeping loudly. On the third day after his death, Navi was buried on the edge of Aston in a coffin made with materials from an old canoe. The house then remained opened for a few more days. A big feast concluded the house of mourning. As time went by, a regular tone of conversation was slowly resumed, and the general busyness of the village was restored. 'Now people shake hands, share food. The sadness is over,' Dinah told me as we were handed big bowls of cooked rice topped with a dollop of tinned corned beef.

Other than food and firewood contributions from other members of the community, no formal material exchanges constituted a part of Paiyamo post-mortem custom. However, the dead child's paternal grandfather, a Sepik man from Oum Three, requested a monetary exchange to honour the tradition of that group. Before Navi's closest family washed, they expected to be given money by the community as means of compensation for their loss. Only once the cash had been received would they conclude the house of mourning. This created a commotion in Paupe. Some of the local women point blank refused to contribute, publicly making all the appropriate excuses ('I would but times are hard and I don't have the money'). They were not happy that the boy's family did not follow the local *kastom*, but when I asked if the community could refuse to give money, I was told: 'No, this would create conflict, gossip, we can't do this.' They also remarked that the number of women in Paupe who were married to the men from the Sepik was so high and their immigration so influential that people from the Sepik can, and do, bring their *kastom* to Paupe and follow it whilst here.

Throughout the week, in the privacy of small gatherings and away from the house where community members mourned the death of Navi, people speculated on what led to the tragic event of the boy's death. Some people said that the blame was with the family of the boy's father, who somehow damaged the child and placed him in harm's way. Or with the father himself – who was known to have had sexual relations with numerous women outside Paupe – and all those who tolerated his uncustomary behaviour. Others said it was Navi's mother's fault for taking him to the bush when she went to wash sago in an area known for the presence of bad *masalai*. 'She should have known better' – they remarked. On a number of occasions comparisons were drawn between Navi and other children whose death, it was suspected, was caused by an act of witchcraft. During pretty much all of those conversations someone would interject and say that it was God himself who decided the time and called Navi to join him in His Kingdom.

The Way to Cargo

I noted above that while I struggled to understand the interplay between what I saw as two distinctive religious domains in Paupe the Paiyamo seemed able to reconcile their traditional and Christian religious registers with much greater ease. Even though they were often aware of the paradoxes in their own narratives and practices, they seemed surprised by the way I approached them. It took me a long time to realise that this was because theorising about Paiyamo beliefs in terms of religious domains inherently created a gap and generated a problem that presented me, not my hosts, with an ontological conundrum. While I focused on what I saw for a long time as conflicting prescriptions of religious doctrine and practice, they experienced their religious lives in a way in which new religious elements dramatically expanded the limits of what could be believed and, therefore, of what might happen. They saw religious knowledge as knowledge with a particular type of effect and enacted their beliefs through a range of religious practices in the hope of eliciting and directing that effect.[23] This is why the names of the *masalai* had to remain hidden and why, every Sunday, in the green church building with its uneven concrete floor and leaky roof, the Paiyamo came together to offer their prayers to God. While the anticipated effects of these enactments continued to address localised concerns such as the quality of social relations, food security, health, welfare and communal wellbeing, the Paiyamo hoped they would also reveal themselves in a novel and still largely alien form of development, modernity and expanded relational networks.

In this context, Christianity held great appeal for the Paiyamo. Unlike the largely localised traditional religions of Papua New Guinea, Christianity is not assigned to a particular territory and entails membership in a community traversing geographical, linguistic and even national confines (see MacDonald 2017). At independence, Papua New Guinea incorporated Christianity in the process of state-building and modernity, and the country's constitution ranked '"Christian principles" with "noble traditions" as values to be cherished and preserved' (Barker 1992: 145). But 'the concept of a deterriorialized Christianity also has the capacity to produce a global citizenry in remote places like Papua New Guinea the members of which feel that, through their practice, they are engaging with the developed outside world' (Jacka 2005: 644).[24] To that end, by partaking in the global phenomenon that is Christianity, the Paiyamo felt able to claim membership in a transnational and transracial system, allowing them, to borrow Robbins' (1998b) phrase, to partake in a future promise extending beyond their current capacities and achievements. The Frieda River Project held a similar appeal for my hosts in Paupe and came with its own set of future promises – of material wealth, political and economic power, access to infrastructure, and opportunities that the Paiyamo knew were 'out there' but that remained stubbornly out of their reach.

Conversion to Christianity was seen by people in Paupe as bringing them a step closer towards that vision. Although they continued to practice elements of their traditional religion, when speaking about communities that did not convert to Christianity, the Paiyamo habitually described them as 'backwards' and 'undeveloped'. Here 'Christianity posed itself as the antithesis of *kastom*. Christian social practices – the light – irradiated dark tradition' (Lindstrom 1993a: 499). They believed and held dear the Christian promise of salvation after death and an everlasting life in heaven, but they also saw conversion to Christianity as a condition for participation in a much earthlier project – one of development and modernity. The way they sometimes spoke about what form they imagined those transformations to take; how they would be both hindered and enabled by elements of local cosmology and *kastom*; and how becoming Christian made them more like the white (developed) people seems to reflect some of Marshall Sahlins' (1992: 13–14) ideas about cultural change in Melanesia. Sahlins used the term 'develop*man*' to describe an indigenous way of coping with capitalism in which, in the face of encroaching Western economy, 'the first commercial impulse of the local people is not to become just like us, but more like themselves' and to reflect on their own ideas of a good life and 'of what mankind is all about'.[25] He argued that as much as encounters with modernity can and do challenge established ideas and practices, they can also reinforce them – encouraging people to revisit and reassess their meaning and value. In an illustration of this process, faced with dramatic changes, communities across Papua New Guinea – Paiyamo included – employed *kastom* as at once an idiom of (national) identity and 'a language of opposition and resistance' (Lindstrom 1993a: 506).

Up until now, I have followed the Paiyamo in their use of the term '*kastom*' without much critical engagement with the term itself. But it is necessary at this stage to point out that far from being straightforward and uncontested, *kastom* is a highly politicised concept in Melanesia, 'a conscious or semi-conscious rectification of the past, and of tradition, made for political purpose' (Leach 2003: 125) and actively embraced both at national and local levels. Counter-intuitively perhaps, *kastom* is a relatively recent notion. The word itself is 'derived from British colonialist terminology via Melanesian Pidgin' (Lindstrom 2008: 166) and can be loosely defined as 'a regional concept of cultural heritage that is connected to global colonial and post-colonial developments' (Otto 2015: 120). It gained importance in Papua New Guinea in the 1960s and 1970s, replacing older phrases such as 'the ways of the past' and 'the ways of the ancestors'. Although the term *kastom* tends to be used in reference to traditional beliefs and practices, its concerns are very much contemporary and serve 'to articulate identity and value *in the present*' (ibid.: 129, emphasis added) often more so than their historical importance. It is 'conspicuous, obvious, and on the table for people to evaluate and debate' (Lindstrom 2008: 165), a political symbol that does not carry meanings; it evokes them (Keesing 1982). To that end, *kastom* can be seen, as argued by Bainton (2010: 177), as 'the product of interpretations of the past by present generations as much as it is the residual essence of the past which has trickled down to the present . . . It is a self-conscious construction of the past which is used to inform present behaviour and a sense of identity'. Understood in those terms, *kastom* is well aligned with Paiyamo ideas about temporalities and the power of finding new meanings and discovering new capacities in old stories and past events to interpret present circumstances and affect the future.

Over the past fifty years, *kastom* has gained significant traction in the formulation of national policy and debate. Unlike the state, which tends to see *kastom* as a placeless political symbol embedded in a national ideology, communities have developed apperceptions of the concept grounded firmly and literally in particular locations (Demian 2015).[26] The Paiyamo clearly differentiated between *kastom* that belongs to them and that which belongs to different people and places. They took pride in the central role *kastom* continued to play in many aspects of their lives and could often be heard commenting that further up the river, in Ok Isai, preoccupation with gold, money and commodities has displaced local *kastom*, making people weaker and less self-sufficient. 'Some of them have already forgotten how to make their gardens. They no longer have their own stick taro and stick banana. They need to look for it elsewhere, and that is not good,' they lamented.[27] In the context of the Frieda River Project, and the landowner ideology, the Paiyamo strategically used elements of their *kastom* in the hope that, if they could find the right mode of engagement with the government, the mining company and, to a lesser extent, the other impact communities, the mine at Frieda River would reveal itself and their lives would be transformed. By engaging with individuals representing the different interest groups of the Project, they enacted a belief in a vision of a prosperous future built on the idea and anticipation of mining-induced development at Frieda.

The Paiyamo commonly described themselves as very remote, located 'too deep inside the bush'. They spoke about feeling forgotten and alienated, deprived of access to goods and services enjoyed by white people and those living in other, already developed parts of Papua New Guinea. The inequalities in wealth they experienced first-hand between themselves and the employees of the company (especially the white, ex-patriate personnel) were striking and further contributed to the local imagery of what their lives on the banks of the Frieda would be like when the mine reveals itself and modernity finally comes. The belief that great wealth will come their way if they can find the right mode of engagement with the elusive mining industry and the people enacting stakeholder roles with regards to the Project is reminiscent of a prominent concept in anthropological debates about Melanesian responses to modernity – cargo cults.[28] At the heart of cargo cults is a promise of a new, prosperous way of life and a 'belief that spiritual and material benefits are obtainable through ritual activity' (Hyndman 1987: 23, see also Lindstrom 1993a). It was loosely based on a premise that if people can crack the code (Worsley 1968) and 'learn the methods used by white people . . . they will also be able to produce wealth' (Lohmann 2001: 102). To give an example, Nick Bainton (2010: 51) described how people on the island of Lihir, now home to the Lihir Gold Mine, attempted to transform their society by first mimicking salient features of administration practices and then those of a business in the hope of finding the 'way to cargo'. In this process, he argued, 'finding the right "formula" or practices became central to Lihirian attempts to understand and control modern processes of production and wealth accumulation, and in each instance, this was inspired by an imagined future that repositioned Lihir itself'. I have not found any evidence of a cargo cult of the kind described by Bainton ever existing at Frieda. But if we take cargo cults to be, as proposed by Nancy McDowell (1988: 122–23), 'an example of how people conceptualise change in the world', if we 'begin by examining the cultural assumptions and frameworks that structure, inform, and give meaning to these so-called cults as well as *other*, similar beliefs and behaviours' and if we focus on ways in which 'a people conceptualise the nature of change and . . . in which change can be effected as well as affected', it would be negligent not to acknowledge the ways that cargoistic thinking informed how the Paiyamo conceptualised the Frieda River Project as a portal through which they could gain access to the material wealth and development they felt they were being denied. They actively partook in the process of planning for a mine at Frieda in the hope that, once operating, the mine would serve as a much desired connection to the world beyond the horizon, fulfilling the long-standing promise of development dating back to the early 1970s when the first mining company began exploration in the area. They expected this change to be great and dramatic and to be revealed to them in the form of the Frieda River Mine.

Conclusions

It is no coincidence that the Paiyamo stories about events of dramatic change tend to include a strong relational component. In most cases, these stories begin with an arrival of a new agent, human or non-human (a *masalai*), in the Frieda area and speak of their encounters with the local communities, as well as material objects and marks in the landscape that they leave behind. Local ideas about relationality, knowledge and personhoods move freely between the visible and the invisible realm in the world of the Paiyamo. In this context, the ways they engaged with *kastom* and Christianity cannot be separated from the ways they approach mineral development at the Frieda River and the wider forces of the globalised world. Through a range of practices, the Paiyamo enacted their belief in the efficacy of various relations, which, they trusted, could create change in their world. This includes relations they built and managed with the company developing the Project. They seemed content to accept that the extent of their transformative qualities would only become apparent under certain conditions and, like most things, may never really be fully known (see chapter 4).

The Paiyamo were very open about their ambitions – both communal and personal. Much of the social life in Paupe was characterised by an ongoing interplay between relational obligations and shared interests and people's, especially young people's, individual aspirations (see chapter 5). They felt left out from the regional and national flows of capital and political representation and longed for change built around prosperity, development and membership in a global community. The way they saw it, new relations could hold the key to knowledge, to revealing the meanings of stories and events and, by shifting perceptions and expanding relational networks, expose in narratives of old new significance and, in some cases, novel sources of leverage and power. They also trusted that certain types of relations made visible through enacted beliefs would create new opportunities for action and bring about a desired effect in the future. However, as pointed out by Crook (2007b: 218), 'relations have to assume a specific form to be effective'. Such a premise marks a point of departure for the next chapter. Learning about the ways the Paiyamo engaged with *kastom* and Christianity can help us understand how they got to know the Frieda River Project and strived to become known to the Project. The parallels are grounded in a local form of engagement with the supernatural, where, to return to Lohmann (2003: 183), the supernatural is modelled as 'a volition that makes things happen, but is not always visible'. Still at exploration stages of development, the Frieda River Project has not yet progressed into a mine and materialised in the form of mining infrastructure and industrial mining activities. However, the anticipated mine at Frieda has already brought new agents into the area, creating new opportunities for potentially transformative relations, and has introduced new ways for people to be. These effects have not all been positive (see chapter 5), but all have altered aspects of the world of the Paiyamo as they know it and experience it.

To highlight the importance of perspective and the role of the supernatural in the world of the Paiyamo, as a final remark let me return one last time to the church at the SBU Camp. One wet and cool Sunday morning, pastors Jacob and John were talking about the importance of religious practice and 'doing things with belief'. They highlighted the need to teach important knowledge and religious practice to the children so that one day they can become good leaders themselves. Pastor John remarked that people need to believe that the mine will bring good things to Frieda. 'If you believe in God, what are you worried about?' he told the gathering. But the last word belonged to pastor Jacob, who turned to the congregation in his loose basketball shorts and a baggy white shirt and said:

> Your big job is to prepare the leaders of tomorrow. Every morning, mothers should pray to the big man in heaven (God) and ask for blessings for her children. All men, too, must go and talk to the big man in the bush (*masalai*) and ask him to look after the children. We do this and then it is going to be alright.

Notes

1. This is far from an exclusive characteristic of the Paiyamo. For another example from the region, see Fraser MacDonald (2017).
2. Lohmann theorised the supernatural as 'a real phenomenon with physical causes and effects' (2003: 176) and defined it as an etic category and an 'ubiqui-

tous mental model that depicts one or more sentient, volitional agencies that are independent of a biological substrate and understood to be the ultimate cause of elements of physical reality' (ibid.: 175).

3. Although Paupe is located in the Ambunti District of the East Sepik Province, the only government services that reached it during my time there were health awareness and assistance projects facilitated and funded by the Telefomin District in Sandaun Province. This introduced an interesting political dynamic among the Paiyamo, many of whom told me that they would like the Sandaun Province to take over official jurisdiction of Paupe. In a period immediately preceding the 2012 national parliamentary elections, some of the houses in Paupe displayed next to each other posters of two candidates – one for Ambunti and one for the Telefomin District – and inhabitants expressed their full intention to vote twice – in Paupe (for the Ambunti seat) as well as in Ok Isai (for the Telefomin seat).
4. In this view, a person 'is a totality, of which any aggregation is but a partial realisation' (Wagner 1991: 172). Wagner claimed that fractality is an inherent characteristic of Melanesian personhood, which 'prevents the differentiation of part from whole, that keeps the imageries of understanding from collapsing into the individuals, groups and categories that constructionism bundles into wholes greater than sums of its parts' (ibid.: 171).
5. Strathern placed her theorising about Melanesian persons in contrast with Western persons, where, LiPuma (1998: 57) argues, the account of Western persons was inadequate and reflective of Western ideology rather than Western notions of personhood. LiPuma criticised Strathern for framing the problematic in terms of a series of dichotomies and hence succumbing to the anthropological tendency to draw comparisons between 'us' and 'others' and focus on difference rather than similarity (LiPuma 1998), where the individual is 'monadic while the dividual is fractal; the individual is atomistic, while the dividual is always socially embedded; the individual is an autonomous social actor, the author of his or her own actions, while the dividual is a heteronomous actor performing a culturally written script; the individual is a free-agent, while the dividual is determined by cultural structures; the individual is egocentric, while the dividual is sociocentric' (Smith 2012: 53). One of the most prominent criticisms of Strathern's model of Melanesian dividuality came from Martha Macintyre (1995), who accused Strathern of romanticising Melanesian worlds, representing them as ahistorical and hypostatic. For Macintyre, such representation denied Melanesia historicity and failed to reflect on the impact of 'a century of interaction with Western ways of thinking about persons and things' (1995: 43).
6. In 2010, Mark Mosko's article 'Partible Penitents: Dividual Personhoods and Christian Practice in Melanesia and the West' challenged the distinction between Melanesian dividual and Christian (Western) individual. Mosko (2010a:

217) argued that Christianity 'also involves detachments and attachments among dividual persons' and that, in practice, individualistic Christian belief and worship 'involves elicitive ritual transactions over parts of dividual persons, divine and human' (ibid.: 218), making it inherently dividual. His arguments were challenged by Robbins (2010), who saw them as demonstrative of anthropologists' tendency to downplay the change that occurs during conversion, and by Barker (2010), who warned that Mosko's analysis diluted Strathern's model to the point where all exchange was taken as dividual exchange (see also Knauft 2010 and Mosko 2010b, 2013 and 2015). The exchange between Mosko and Robbins triggered a broader discussion, which was presented in a special section of *HAU: Journal of Ethnographic Theory* (2015, vol. 5:1). Introducing the section, Bialecki and Daswani (2015: 287) described it as a collection of essays that ask 'whether it is possible to break down the in/dividualist distinction and think it anew, and thereby apprehend the forms in which new modes of interrelatedness might appear to us'. The collection advocates a shift of focus away from the dichotomy and towards the study of 'the processes that allow a range of variations in modes of relation and personhood that would encompass both of these hypostases' (ibid.: 285).

7. See Jorgensen (2014) for a version of this process among the Min people affected by the Ok Tedi mine.
8. In local vernacular, *sinapi* means crocodile, while *Mele* signifies the river Melegeme.
9. Awageme is a Paiyamo Amoke name and Amosai a Telefomin name. Inclusion of both names in the story is indicative of the dynamics of local politics.
10. *Mumu* is a type of oven built from stones heated in the fire. Food is placed onto banana leaves covering a layer of hot red stones. The food is then covered by another layer of leaves and sealed with another layer of stones.
11. As one of my interlocutors explained to me, another man knowing the name of the *masalai* could come to Nena junction and impersonate a Melefale man. He could then try to use the powers of the *masalai* against the Melefale, which is why it was so important to keep that name hidden.
12. Although Peter did provide the secret name of Melesinapi, I chose not to reveal it here.
13. For examples from the Min area, see Barth (1975, 1987); Brumbaugh (1980, 1990); Crook (1999); Jones (1980); and Jorgensen (1981, 1990a). For insightful accounts from other parts of the country, see, for example, Harrison (1995); Keesing (1987); Lattas (1998); Lindstrom (1984, 1990); and Wagner (1984), among many others.
14. The sacred knowledge, the Baktaman believed, was received from the now deceased ancestors and, in Barth's view, the ritualistic initiations 'provided the validating organisational form in which [it] was reproduced and transmitted' (2002: 4). The ethnographer used an image of Chinese boxes to illustrate the

structure of Baktaman religious knowledge and suggest that, from the perspective of an initiate at least, 'it is constructed with multiple levels, and each level is organised as to obscure the next level' (1975: 218). He argued that the information obtained at every consecutive stage of initiation may be viewed in terms of a specific right gained (1975) and that the value of knowledge was 'enhanced by veiling it and sharing with as few as possible' (1990: 641). At the same time, he wrote, the total corpus of Baktaman knowledge could only persist 'to the extent that its parts are frequently re-created as messages and thereby transmitted' (1975: 255).

15. Jorgensen (1990a: 41) elaborates further: '"Turning" (*fukpela kupkamin*) in Telefol includes the notion of translating and transforming, and it is worth noting that the latter sense encompasses our idea of magic. Taking all these senses together, we can say that "turning" is what Telefol initiations are all about'.
16. Engaging critically with Barth's model of Min secrecy, which places at its core an intricate system of male initiations, Crook proposed an alternative approach to eliciting understanding of Min knowledge practices. He interpreted *awem* as important rather than sacred or secret knowledge and claimed that it emerged from and created social ties and defined the Min for who they were (1999, 2007b, 2009). His writing shifted the emphasis from the significance of a secret for the Min to the role of acquiring important knowledge in constructing personhoods and managing a range of social relations that extend beyond the sacred cult houses in the region.
17. See also Gardner (1981); Jorgensen (2005, 2007b); Lohmann (2000, 2007); Morren (1986); and Robbins (2001, 2007).
18. Note that the Oksapmin were not directly affected by Diyos' religious revival and had been first evangelised some years earlier by the Australian Baptist Missionary Society based in Telefomin (where the Society established itself in 1952), and then from a mission station on Oksapmin territory that opened in 1962 in the lower Tekin Valley. MacDonald writes: 'The revival did come to Oksapmin, via the Bimin-Kuskusmin, around 1980 and, while intense and widespread, it didn't produce the same kinds of radical change seen in other societies in the region, owing primarily to the fact that these transformations had been taking place in a more moderate manner within Oksapmin society for over a decade. But while the Oksapmin version of the revival cannot be credited with catalysing popular conversion, it did institute charismatic techniques of worship, such as speaking in tongues, shaking, and healing' (2013: 82, see also MacDonald, 2014).
19. The reference to *masalai* as 'Satan' or 'sin' can be found in a number of accounts from different part of PNG. For example, Jerry Jacka's (2010: 38) account of Christianity, ecology and resource management in Porgera (Enga Province) describes how local Christians were able to talk about traditional beliefs 'by linking them to Satan'. In doing so, they did not 'deny the reality of the spirits;

rather, it is that their basic nature has been misinterpreted by Porgerans who were not aware of Christianity'.

20. Washing sago is the process of extracting starch from the pith of the trunk of a Metroxylon palm.
21. I have heard two different explanations for what happened to the bones of people who did not hold a prominent status in the community. One was that their bones were simply left in the bush; the other was that families collected those bones, put them in *bilums* and hung them up on the outside walls of their houses. Only the bones of the greatest warriors and notable members of the *kastom line* would be placed in the cult house.
22. Following the introduction of what people in Paupe referred to as 'the ways of white people', the local practice of leaving bodies to decompose deep in the rainforest, away from the village, was abandoned. They now buried the bodies of their dead in the ground, either on the edge of the village or, in rare and isolated cases, underneath the deceased's house, in which case the house itself became the grave, left uninhabited and allowed to deteriorate.
23. See also Robbins (2001: 906) on how 'the Urapmin understand all rituals to be discrete, formal events of this kind. They also understand them to commit those who participate in them to the goals at which they are aimed. This commitment is grounded in participation itself and is not a function of the intentions of participants. One cannot pray for something without committing to wanting it to happen'.
24. Jacka's (2005: 649) comment relates to his work among the Ipili affected by the Porgera gold and copper mine. The Ipili used the concept of 'heaven' to reterritorialise otherwise placeless Christianity. Jacka writes: 'The Ipili have embraced millennialism and their version of "heaven" as a means to resolve the contradictions of social relations being disembedded and distanciated in an era of globalisation and development. It is precisely the idea that heaven, as the final place, will fulfil the goal of creating new identities and social relations that are currently being denied to the Ipili'.
25. At the core of Sahlins' (1992: 24) model is a premise that 'humiliation is an important stage of economic development, a necessary condition for economic "takeoff"' and that 'in order to desire the benefits of "progress", its material wonders and comforts, all indigenous senses of worth, both the people's self-worth and the value of their objects have to be depreciated'. The model served as a point of departure for a noteworthy volume *The Making of Global and Local Modernities in Melanesia: Humiliation, Transformation and the Nature of Cultural Change*, edited by Joel Robbins and Holly Wardlow (2005).
26. Demian (2015: 95) drew a clear line between custom (in the legal sense) and *kastom* (in the ideological sense) and argued that 'in that Papua New Guinean apperceptions of kastom are grounded (again literally) in particular locations, which stand in particular relations to the groups of people inhabiting them . . .

the legal life of custom in Papua New Guinea has taken it so far away from any relationship to place as to have become, in essence, placeless'.

27. In a similar vein, Leach (2003: 131) remarked that people he worked with on the Rai Coast 'are distinctly proud of their adherence to and continuing knowledge of *kastom*. They speak with scorn when their neighbours, who have given up *kastom* in favour of seeking power through Church or business, come to them for assistance with divination or love magic'.
28. In very general terms, in pre-independence Papua New Guinea, the phenomenon of cargo cults was based on indigenous people's 'belief that European goods (cargo) – ships, aircraft, trade articles, and military equipment – are not man-made but have to be obtained from a non-human or divine source. It expresses its followers' dissatisfaction with their status in colonial society, which is to be improved imminently or eventually by the acquisition of new wealth' (Lawrence 1964: 1). Some of the best-known studies of cargo cults in Melanesia include: Kenelm Burridge (1960) *Mambu: A Melanesian Millennium*; Peter Lawrence's (1964) *Road Belong Cargo: A Study of the Cargo Movement in the Southern Madang District New Guinea*; Peter Worsley (1968) *The Trumpet Shall Sound: A Study of 'Cargo' Cults in Melanesia*; Lamont Lindstrom (1993b) *Cargo Cult: Strange Stories of Desire from Melanesia and Beyond*; and Andrew Lattas (1998) *Cultures of Secrecy: Reinventing Race in Bush Kaliai Cargo Cults*.

Illustration 4.1. View of the Frieda River from Dinah's banana garden above the SBU Camp. Photograph by the author.

4

The Book of Paiyamo

Introduction

I spent a lot of my time at Frieda talking to people about land. I enjoyed the intimacy with which people in Paupe talked about their places. Conversations about land inevitably led to ancestral stories and biographical narratives, descriptions of sacred and other significant sites, funny anecdotes and portrayals of *kastom* as well as dramatic events in the life of the Paupe community and its people. It was clear to me from those encounters that land and places were not just a setting for the events and stories told but active and crucially important parts of the complex network of dynamic relations that make up the world of the Paiyamo. What was also clear was the way in which the prospect of a mining development in the area introduced new forms of relations between people and land that challenged a whole array of local ideas and left many of my interlocutors both conflicted and confused.

During one such conversation, pastor Jacob told me about a man in Brisbane who was helping him put together a document, for the company and the government, with information on the clan structure of the Paiyamo Sane and the boundaries of land owned by individual clans. 'We do not need to record the land and groups for the Paiyamo. The Paiyamo lineages know the boundaries of the land and there is no dispute between them. But the document will make the Paiyamo stronger for the company and the government,' he told me, standing outside Dinah's house in a worn T-shirt hanging over a pair of baggy shorts, with a notebook in one hand and a large bush knife in the other. Minutes later, he looked at me with a glimmer in his eyes and added: 'A man in Brisbane is putting the document of the Paiyamo land boundaries together. But you know it already, you have seen it.' It was not until I actually saw the finished document that I remembered the loose pieces of paper the pastor had shown me during our early conversation about Paiyamo lineages, stories and *masalai* (see chapter 3) and realised that I had indeed already seen it but did not know it.

The document was entitled 'Paiyamo Sane: Clans, Histories and their Make Up', although it was more commonly referred to as 'The Big Book' or simply as 'The Book'. The 'man in Brisbane' was a son of an Imlou woman from Duranmin and a Telefol man from Eliptamin. He was brought up in Duranmin but lived in Australia, where he worked for Xstrata Copper Ltd's Geographical Information Systems (GIS) division. When I first met him, he told me that The Book was his idea. He said that he had been watching pastor Jacob closely and that he felt that the pastor was not able to lead the community in a way that would fulfil the demands of the Project. Having worked for the company in Australia, and having experienced the Western system of formal education, he believed that he had the knowledge and access to the kinds of networks that could award weight and efficacy to the claims the Paiyamo made of their land and of the Frieda River Project. He referred to the Paiyamo as 'his people' and said that he wanted to help them as they prepared for an operating mine at Frieda. This chapter uses a metanarrative of The Book to introduce ways in which encounters with the company developing the Project introduced the Paiyamo to new ways of 'doing knowledge' and making it visible. It pulls together themes of knowledge and its forms and exposes the changing politics of land playing out at Frieda, driven by the process of land commodification, which was not yet fully realised but was already happening. The chapter describes how the context of the Frieda River Project introduced new ways of thinking and talking about the land within an externally imposed system of classification and how people made sense of it in the local cosmological setting. The effects of these processes have already started to show, not just in terms of the anticipated and strategic strengthening of claims and positions in relation to the Project. The chapter describes how the changing nature of land boundaries and new forms of delineation, originally introduced for an externally oriented purpose, started to become internalised, introducing new forms of inclusion and exclusion in Paupe and creating tensions within the community.

The Book

Looking at it lying on the floor in Dinah's house, 46 pages bound together with a narrow, black spiral coil, The Book did not look particularly grand. But its description as The Big Book had little to do with its size and everything to do with its content and the excitement and expectations it elicited among my hosts in Paupe. A small, simple graphic decorated its cover page – a river bend winding around the back of two mountains known locally as Arimoku and Biamaku. This graphic was expected to become a business logo for the Paiyamo Sane Grup [*sic*] of Telefomin and Ambunti Districts, who were acknowledged collectively as the authors of the document. The Book described the group as 'distinct in the language it uses . . . only known to the people that dwell within the vicinity set by the great ancestor (Melesinapi) and the allied communities or clans' (PSGTAD 2011: 3). It defined Paiyamo Sane as 'composed of the mountaineers – the Yuupaiyamos and the coastal or river bank dwellers – the Suupaiyamos' and claimed that 'their distinction and unity is indicated by their language, the [shared] culture and tradition, the land boundaries, historical war [alliances] and historical trade ties' (ibid.). The Book began with the following preface (original wording and spelling have been maintained in all quotes from the document):

> As always is a Melanesian way, stories, legends and family ties are passed on orally from fathers to sons, generations after generations. It is at this time that the group (Paiyamo Sane) had decided to have what has always stayed in their minds on paper. The purpose of the paper is to layout and expose our identity, relationships, the land boundaries and the family ties (family trees) for the ease of future generations' usage, for whoever that is interested in our land vicinity for any developments and for the general public to know and respect our inhabitants. In order to have what we always have in minds on paper, so many leaders, elders and all members of each clans of the Paiyamo Sane Grup were consulted and the result (this paper) is a collaborative effort. As such, the contributors deserve credit and recognition for their contribution in stories, time in identifying their land or territorial boundaries and their family trees and family name listings and the contribution in kind for the execution of this group. Though all or many have taken part in one way or another in making this paper a success, it is impossible to adequately acknowledge all. By naming only a few here, the group extends its appreciation to all members of each clan. (PSGTAD 2011: 2)

The idea for The Book and the proposed content of the document was first discussed during a meeting held in Paupe in March 2011. There, leaders of the different clans acknowledged the need to present the Paiyamo land boundaries in a form mimicking the way in which the company itself identified and recorded the owners of the lands in the Project area. They hoped that if they made their knowledge of the land visible to the company in a format their representatives were familiar with, the company would be able to respond to and act on that knowledge. During the meeting, the Imlou clan was officially recognised as the sixth Paiyamo clan, and the Imaka-Kuwari clan from Tau gained recognition as a sub-clan of the Melefale. At the same time, two other groups recognised as members of the Paiyamo Sane during the group's previous meeting (held in September 2009) were stripped of their status and proclaimed to be 'outside too much' to be included in the Paiyamo collective. The leaders made a decision that The Book should contain what they referred to as the Paiyamo Sane Grup Clan Structure (see Figure

4.1.) alongside what were described as the family trees of each of the six distinctive clans: Melefale, Imlou, Nesiar, Suna, Inaku and Ayamo, and lists of their current members. As reflected in the meeting notes, the leaders agreed to work collaboratively to identify the land boundaries between the clans, which were later captured on GIS maps and included in the document. Minutes from the March meeting include a comment made by Moris, leader of the Imlou clan. Although the content of the document was described as factual, whilst negotiating the ways in which the Paiyamo clan structures are represented in The Book, Moris proclaimed: 'The clan structure will not be like this forever. No. Suppose the company will start working on the ground marked as belonging to one of the clans, we will change the structure. At that time we will be able to change our leaders, too' (PSGTAD 2011, Appendix B: 4). This comment reflects a concern raised by several Paiyamo leaders in the process of preparing the document. For generations, local histories at Frieda were recorded and reproduced via an oral tradition. This made them highly mouldable to present circumstances and future-oriented aspirations. They evolved over time as new elements and interpretations embedded in contemporary issues and the dynamics of local politics were brought to light. Once written down, the leaders feared that such stories only reflected one of many possible and constantly evolving narratives and awarded individuals with powers that changeable oral traditions kept in check. This may explain why the document concluded with the following statement: 'Finally, the geography, maps and family listings provided in this paper are a collaborative effort. However, we can't claim this paper to be final but are open for criticism and comments hence have room for improvement and amendment' (PSGTAD 2011: 13).

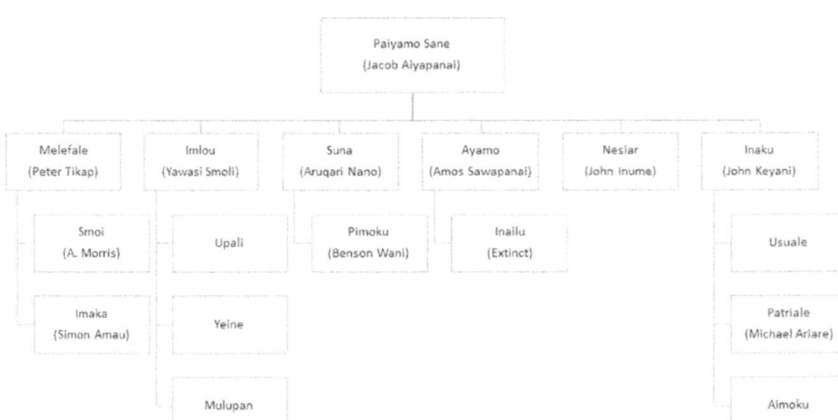

Figure 4.1. Paiyamo Sane Grup Clan Structure and Leadership (adapted from PSGTAD 2011: 5). Published with permission.

Alongside information about clan structures and membership, The Book offered an extensive version of the myth of Melesinapi, which was unveiled during the March meeting.[1] The myth told the story of 'how Melefale clan inhabits land extending from the mountains to the coast or low lands' (PSGTAD 2011: 7) and served as evidence attesting to the shared ancestry of the Melefale and Imlou clans. The version of the myth named 'the groups that shared the common borders and boundaries with the great legendary father' (ibid.: 8) and offered a detailed description of the *masalai* travels throughout the land, naming a number of geographical features – rivers, creeks and mountains – and thus identifying land boundaries as marked out by Melesinapi himself on his journeys around the area. The true name of the *masalai* was strategically hidden from the publication. In its place, the word 'named' was used in the narrative's opening statement: 'Paiyamo Sane legendary father, Melesinapi – crocodile (*named*) and his brother Anau lived in Kapkapu amsilima' (ibid.). The story in the document described how Melesinapi asked his son to 'go back and look after the land and boundaries but become visible and be a human being' (ibid.: 10):

> The great legendary father, while still a crocodile used to travel his boundaries. He goes up and down Neiligeme, Melegeme (Melia River) and goes past Benegeme (Ok Binai River) and up to Sialema (Smomilia). He (Melesinapi) still looks after his boundary by travelling the rivers. During the stay, he laid four eggs. Three of them hatched and became his sons however; they were spirits – *their names held back*. He told them to live invisible and to look after the land and boundaries that extend from Mlekame junction down to Arimoku and Biamaku. The last egg hatched and also became a son [Obilebo]. He told him to go visible and be in the human form. He is the founder of Melefale especially the Imaka/Smoi. He told his son that he will not go anyway but stay and look after them. (Ibid.)

Turning its attention to the Suupaiyamo, the document noted that unlike the Yuupaiyamo (Melefale and Imlou) they did not have a shared ancestry. Instead, each of the four Suupaiyamo clans (Nesiar, Suna, Inaku and Ayamo) had its own ancestral history. When I asked the man who produced the document why their stories were not presented, he told me that the leaders of those clans failed to provide their stories in time. Some time later, Peter explained to me that there were currently two pieces of company infrastructure on the Paiyamo land – a communications tower on top of the Arimoku Mountain, and the Frieda River Airstrip on Piamoku. Both these sites were located on Melefale land. It was also their land that bordered the contested Miyanmin and Telefomin territories. As the company only used the Melefale land, the document was biased in their favour. 'Suppose in the future, when plans are made for construction of a road and a pipeline on other clans' land, those clans will then present their stories, too,' Peter added. Although it was clear that he was able to work with some clans more closely than others, the 'man in Brisbane' who produced the document continued to assert that the document was a truly collaborative effort of all the Paiyamo and that it was a book 'which only showed facts'.

Another key feature of the document was that it used Paiyamo Amoke terms to name geographical features in the area. During the time I spent with them, my hosts in Paupe would often use local topographical nomenclature to support claims of rights to land at Frieda. In The Book they collectively stated that:

> We pride ourselves in the naming and originality of names we have for the rivers, creeks, features, mountains and ghosts. Though our younger generations are inventing new names, everything on and within the vicinity of our land is in our traditional dialect only known to us. Some people from outside our boundary had come in and invented names of the rivers; Example, Amosai (Telefol)- Wakame (Paiyamo), but we doubt if they know the hidden names and ghosts in these areas. Our clear land territories and demarcations tie to our histories and the language we speak defines our land extent and community ties. Our land identity and the history of each clan is known and respected. Each and everyone within this group – Paiyamo Sane – knows and respects each other's land boundaries. We also know clearly whom we originally shared common borders with and treat as brothers and whom we treat as enemies – traditionally (the communities out of the Paiyamo Sane Grup territory). (PSGTAD 2011: 13)

Presentation of The Book to the company officials was planned for early November. The evening before the event was due to take place, pastor Jacob called a community meeting. The purpose of the get-together, he told me, was for the people of Paupe to see and confirm The Book and the knowledge it represented before it was handed over to the Community Affairs officers. The response to pastor Jacob's call was good, and the turnout to the gathering was high. Men (and some women) from Aston and Aslimbum joined the SBU Camp's inhabitants outside Dinah's house to listen to pastor Jacob and, as many of them told me, to see The Book for the first time. Opening the meeting, the pastor announced with pride: 'Before, the genealogies were not possible, as there were no educated men. Now, a Paiyamo man from Duranmin, now based in Brisbane, is preparing the document. He is an educated man, and he knows how to work a computer. The paperwork could cost us big money, K13 000. But I am his uncle, so he does it for free.'[2]

The people who partook in the lively gathering were encouraged to look at and comment on the document. Talks and discussions outside my window continued until late that night. People seemed intrigued by The Book. They tried to find their own names in its pages and exclaimed joyfully when they finally saw them written down in sharp black letters on white paper.

At different points of the evening, a number of individuals remarked loudly that the document was demonstrative of the Paiyamo's new capacities in engaging with the company and the government. They expressed pride in the fact that the document was written by a Paiyamo man and that the group was able to put it together without the involvement of external consultants often used by the company. They told me that this was the first time they had tried to record this information themselves and took pride in the fact that they were able to replicate in The Book a format of documents normally produced by the company and the government. They drew comparisons between themselves, the Telefols and the Miyans and remarked gladly that 'the Telefols and the Miyans have many important people who went to school and some even work in cities. But it is us, the Paiyamo, who are first to produce such document'.

Amongst the general excitement, my attention was drawn to two aspects of The Book that, to me, made this event of community consultation really quite unique and peculiar. One, the majority of the document was written in English – a language largely absent from Paupe and understood to a greater or lesser extent only by very few individuals in the village. Pastor Jacob himself had no command of this language and was therefore unable to read significant parts of the document, including the all-important story of Melesinapi. Two, the stated aim of the meeting was to present the final draft of The Book to the Paupe's Paiyamo and to ask them to check the document, identify any inaccuracies and agree on its final form. But by the time the meeting took place, The Book had already been printed and bound in the company camp, ready for presentation to the Community Affairs personnel the following day. As such, at that stage, any alterations to the content of the document were no longer possible. No one really seemed to mind, though. This is because confirming the content of the book and the information presented in its pages was never really the purpose of the showing. The Paiyamo gathered outside the house that night saw The Book as a happening rather than an object, a moment of unity through which they endorsed the document. That moment revealed to them new capacities to enact and present their knowledge in a way that, they hoped, would 'make them stronger' in the context of the Frieda River Project.

The following morning, pastor Jacob got up early. His pale blue shirt with long sleeves and a crinkled collar was tucked carefully into his long combat trousers. He liked to dress in a particular way when preparing for a meeting with the company or government representatives. On those occasions, instead of wearing one of his usual outfits consisting of a T-shirt over basketball shorts, he opted for what he himself would often describe as 'a white man's clothes' – one of the two shirts that he owned and kept in a locked room in Dinah's house, hanging away from the damp walls, and a pair of long trousers or thick black combat shorts. The pastor also owned a tie, which he would wear to what he considered to be the most important of meetings but which could more commonly be spotted serving as a belt, tied securely around his narrow waist. After a cup of weak and very sweet coffee, he sat down on the steps of Dinah's house and remained there, sitting still, with a copy of The Book in a large brown envelope resting on his lap. At the sound of a motorised canoe approaching the waterfront outside the SBU Camp, he got up without haste and slowly walked over to the steep bank of the Frieda River where, outside the old, dilapidated clinic building four members of the Community Affairs team were disembarking the canoe and climbing their way up to the flat grass of the riverbank.

The ceremony that followed was brief, witnessed by relatively few members of the Paupe community (mostly Melefale) – significantly fewer than were present during the community gathering held at the SBU Camp the evening before. Mirrored on the model characterising Project meetings arranged by the company, it was opened by pastor Jacob, who welcomed the company visitors and thanked them for their presence. Before presenting a copy of the document to the company's representatives, he reminded them that, unlike their neighbours, the Paiyamo clans knew the boundaries of their land and that there were no land disputes among them. He presented The Book with great pride and, echoing some of the comments made by community members the evening before, remarked on at least two occasions that neither the Telefomin nor the Miyanmin communities have presented their genealogies this way. The Paiyamo were the first impact community of the Frieda River Project to collectively document in a written form, for the company and the government, their genealogies and their landownership structures themselves. 'This book is really important,' he said, in a loud, clear voice as he passed a copy of the document to the Community Affairs officials. Members of the Community Affairs team listened attentively and accepted The Book with expressions of gratitude and respect. In a brief response to the pastor, they acknowledged the importance and magnitude of the task and thanked the Paiyamo for their hard work. Before boarding the motorised canoe and returning to the company camp, they once again complimented the Paiyamo on the document and said that it was going to be very useful for the company, and for the Project. Shortly after they left the waterfront, pastor Jacob too boarded his canoe and returned to Seligeme, carrying under his arm banana rhizomes to plant in his garden. He was tired but seemed pleased with the way The Book had been received both by members of the community and the Community Affairs team. As the canoe started to move away from the clay embankment, he waved at Dinah, Mawi and I as we stood on the riverbank to see him off. He then fixed his eyes on the river currents ahead and, skilfully steering the engine mounted at the back of his dugout canoe, sped away.

Illustration 4.2. Pastor Jacob Aiyapanai. Photograph by the author.

Knowing Land

My interlocutors in Paupe would often make evident their detailed and intricate knowledge of local topography and describe the role it played in their biographical and communal histories. They drew this knowledge from ancestral myths and personal stories, in which vivid descriptions of the landscape were intricately interwoven into narratives told about people, *masalai*, places and happenings. In the earlier chapters, I noted that the Paiyamo used myths to tell their stories but also to elicit their multiple histories and to explain happenings and relations that made them who they are. I argued there, and I want to repeat this argument here, that the Paiyamo narratives of past events cannot be detached from their contemporary contexts. What is more, those narratives often feed into local imagery of the future, uncovering an array of desires and aspirations. This is by no means unique to the Paiyamo. Myths have always functioned as a mode of communication and of building relations in one way or another, shifting in response to needs and circumstances (Sahlins 1999). Literature is filled with examples of how, in the light of changing conditions, local myths become globalised (e.g. Lattas 2005); global stories localised (e.g. Jorgensen 2014); and how old narratives are reimagined (e.g. Macdonald 2017) and 'retrospectively given a weighted significance that [they] previously did not have' (Strathern and Stewart 1998b: 211).

Like all important *kastom* knowledge, the Paiyamo traced their most powerful narratives, like the story of Melesinapi, to the *haus tambaran* and to a time long ago when their first ancestors travelled the land marking the boundaries of the Paiyamo lands. More often than not, when talking about their land and telling their ancestral narratives, the Paiyamo would make a clear distinction between their own 'true, original and authentic' stories passed down from generation to generation and the 'new, made up stories' that emerged for the first time only very recently in the context of the Frieda River Project – even though all such comments should be read and interpreted in the context of competing for recognition and access to anticipated benefits from the Project. 'The Telefols, they are creating new stories – they just want to get the money,' my Paupe hosts would often tell me. 'They talk with the company. But some of those stories, we have never heard before,' they would say. Then they would often add that 'the Telefols and Miyans – they are disputing their land boundaries. New stories keep coming up. But not the Paiyamo. Here, the boundaries are not marks which belong to the present – the boundaries were marked in the days before our ancestors and there is no dispute here. We know our land and our stories'.[3]

It is no coincidence that, together with maps, family trees and clan listings, The Book of Paiyamo contained within its pages a version of the story of Melesinapi and a detailed description of his movements around the area, thus emplacing the myth in the local topography. The Book named several of what my hosts in Paupe considered to be sacred sites, some of which were pointed out to me as I accompanied them on their travels up and down the

Frieda River. Jorgensen (1990b: 54) wrote that such localisation of stories in the known landscape transformed people's experience of land and awarded the myths 'an indispensable bridgehead in the everyday world'. Further, he argued that the myths' presence in particular places awarded them the realism of reminiscences; 'to insist that mythic events took place *here*, where we stand, is to insist that they are in some sense part of our world' (ibid.: 51) and to vouch for their authenticity. By placing the myth of Melesinapi in known places, the Paiyamo were able not only to assert that the *masalai* was a part of their world but were also able to use their relationship with the *masalai* to claim rights to particular areas of land – a tactic they readily used to claim a position of landowners in the Project area. For them too, like for the Telefomin studied by Jorgensen (ibid.: 48), 'sacred places [were] not only places where things happened: they [were] also places where things continue[d] to happen', and the telling and retelling of ancestral myths allowed the past to enter the present, where it could then be reshaped and remodelled.

When talking about their knowledge of the land, alongside myths and biographical stories, the Paiyamo would often draw attention to the importance of ancestral songs sang during communal *singsing* events. Like ancestral narratives, these songs were passed on between generations and constituted an important body of knowledge for the Paiyamo. 'The Paiyamo land is very big,' one of the village elders explained to me. 'The songs teach people about places and boundaries so that they know where the Paiyamo land is. Some places are so far away that people might never see them. But if they know the songs they will be able to identify them when they do.' The Paiyamo told me that when Jim Fingleton, a consultant hired by the company to assist in the task of identifying the Project's landowners, arrived at Frieda, they used the lyrics of these songs to describe to him their relationship with particular areas of land and, what follows, the complex relations between the different Paiyamo lineages and their neighbours, as reflected in local topography. During his visit, Dr Jim, as they liked to call him, worked in the old Community Affairs office at the Airstrip Camp. I caught a few glimpses of him at work, bent over a large table, closely studying large maps of the local area. My Paupe hosts said later that, upon hearing their myths and songs, Dr Jim took a pen, put it to the map and marked extensive grounds stretching along the banks of the Frieda River as land which, different Paiyamo groups claimed, belonged to them. Generally, people were happy to share their knowledge of their land with the consultant and were glad to see it marked out on paper and made visible. As the time went by, however, they grew increasingly disappointed and frustrated with the process as the anticipated results failed to materialise and the company continued to ask questions and request further information about the Paiyamo ownership of their lands. 'Dr Jim was good. But it did not work. We told the company everything and they keep asking questions. But we said it already. They should have listened to us,' one of my local interlocutors remarked with resignation.

Each of the Paiyamo groups living in Paupe had its own repertoire of songs that described what, at the time of my fieldwork, they habitually called 'the ground that belongs to us'. The songs offered an emplaced account of their history and served as evidence of their relationship with given sites and territories. Traditionally, they were only ever sang publicly during *singsing* events, lasting from sunset to sunrise. Unlike many of their neighbours, the Paiyamo held all their *singsing* events indoors rather than outdoors. 'If you sing outside, your body will age overnight; you will become an old person, and then you will die,' they would tell me, although sometimes the explanations were much more pragmatic. 'During the event we wear big feather headdress. What if it rains? If we are outside, the rain will wet the feathers and the headdress will be ruined,' as I heard one of the local men say. In the time of lineage houses, *singsing* gatherings offered rare and eagerly anticipated opportunities for the Paiyamo Sane groups to get together. They would normally be called by one of the longhouse communities, which then had to provide food, firewood and shelter for its guests. All other lineages had to be invited to the gathering. The songs could only ever be sang in the presence of all groups, to ensure than none of the lineages claimed another's land by changing the lyrics of their songs. Although by the time of my fieldwork, longhouse communities were a thing of the past, *singsing* events were still occasionally held in Paupe, and the lyrics of the songs continued to serve as an archive of local knowledge of land and territory, passed on from fathers to increasingly disinterested sons. Here is an extract of lyrics from an ancestral song traditionally performed during a *singsing*, as sang to me by a Nesiar leader, Jonah:

Liliubo liliubo.	I am ready to go now, I go now.
Yupane mawiaro lili.	To the Yupane river sun I go.
Lue mawialo lili.	To the Lue river sun I go.
Asinebale mawialo lili.	To the Asienebale river sun I go.
Liliubo, lili.	I am ready to go, I go.

In the previous chapter, I described how the Paiyamo saw their religious knowledge as knowledge with a particular type of effect and how they enacted a range of beliefs in the hope of eliciting and directing this effect. Here I would like to expand on this perspective and apply it to knowledge more broadly, following Strathern (2005) and Leach (2012a, 2012b) in viewing knowledge as information capable of causing effect in people and places. Seen in those terms, knowledge becomes deeply relational. It requires a knower to make it effective, and its effect is tied to the quality of social relations and dependent on the knower's ability to make the knowledge (and the relations) visible in a specific form. In many ways, it also becomes localised – belonging to a place as well as a people. The Paiyamo would often speak about knowledge that belonged to them and their places, contrasting it with knowledge that belonged to elsewhere and to someone else. They were intrigued by the ways in which the company and the government 'did knowledge' and the systems (legal, economic, bureaucratic) that awarded this knowledge with effect. They would commonly describe it as 'knowledge that is not from here' and approach it with caution but also with great curiosity.

Using a repertoire of emplaced myths and songs, the Paiyamo were able to evoke great and detailed knowledge of their land and land boundaries, acting upon and making visible their histories and relations. But they also talked about how, over time, they realised that new types of markings and new ways of demonstrating their relationship with the land were needed if their knowledge of the land was to have power and the desired effect in the context of the Frieda River Project. As an impact community actively partaking in stakeholder engagements at the Project, they became increasingly aware of new forms of presenting knowledge, what Joshua Bell (2014: 106) termed 'veracity of form' and defined as ways in which communities 'evaluate the forms that knowledge takes, the discourses that emerge around them and the ways in which the materiality of these forms effects these engagements'.[4] 'The songs don't change, they stay the same,' Tikap once told me as we sat in a wind house, sheltering ourselves from the heat of the midday sun. 'They are documents we have and for us the myths and songs are enough. All clans know this and respect the land boundaries between them,' he continued. 'Songs will remain true and people will still sing them. But we now also need a map of our lands because of the company and the government,' added Vincent, Tikap's grandson, joining in the conversation.[5]

Land Happens

At the time of my fieldwork, the Paiyamo already had a long history of partaking in the process of preparing for an operating mine at Frieda. In order to perform their ascribed stakeholder role and gain access to a relational nexus that bore promise of opportunities reaching far beyond the Frieda River valley and that could, one day, allow them to travel the globe, they tied themselves to specific people and places and revealed themselves to the company as clearly delineated landowning clans. But in understanding the Paiyamo as landowners in the Frieda River Project area, it is crucial to realise that, as argued by Leach (2004: 43), in Papua New Guinea '[l]and is not owned out of context, out of history, and out of particular relations that make it valuable'. All of these factors became internalised in the way the Paiyamo conceived of land ownership and adopted the landowner label. In the pages of The Book, the clans claimed to represent ancestral and interpersonal relations among the Paiyamo. But they were equally indicative of an introduced categorical form of relations, prescribed by the form of a stakeholder dialogue and embedded firmly in the specific roles expected of the different stakeholders in a process of planning for a large-scale resource extraction project (see chapter 1). These roles were enabled by a bureaucratic apparatus of recognition that required information about the people to be presented in particular ways. But, as noted by Monica Minnegal et al. (2015: 496), 'where people commit expressions of social form to paper they inevitably reshape previously operative social and political forms', not least because they are 'focused on the definition of categories and drawing of boundaries rather than on the negotiation of relations' (ibid.: 500). It is because such documents are, to a large extent, creative rather than reflective elicitations of relations. Their seemingly timeless materiality, as observed by Bell (2009: 35), 'constraints and shapes the ways in which they are generative of social relations, elicit the responses they do, and instantiate particular regimes of value and knowledge'.

Following independence, and fuelled by the mineral prospecting boom of the early 1980s, the Papua New Guinean government collectively identified Papua New Guineans as 'customary landowners', promoting an 'ideology of "landownership" which is both a form of local custom and a form of national identity' (Filer 1997a: 156). The government developed a series of legislative instruments based on what Leach (2014: 55–56) described as a default assumption about 'property rights being the enforceable and visible manifestation of people's connection to land' that were 'explicit about making it possible to transform complex and embedded, distributed, and fragmented ownership into a single property right that can be transacted'. The government's hope was that these instruments would provide business opportunities for rural Papua New Guineans (Power 2008) and facilitate what Stead (2017: 48) described as a process of '"mobilization" of customary land for "development"', without alienating it from its traditional owners (see also Weiner and Glaskin 2007). However, in large parts of the country, 'one does not own land as an object, but rather the rights (bundle of rights) of secured access, use and enjoyment for a determinable, usually long-term period' (Kalinoe 2004: 60). It soon became apparent than in those places people's 'attitudes to land, with their claims to use rights, but not ownership and disposal rights, by virtue of acknowledged connections to kin-defined, exchange-validated, and locally constituted corporations, contrast with notions of firmly and often legalistically defined rights to land areas with clearly determined boundaries, to which the rights holders claim freehold use and permanent title' (Sillitoe 1999: 332). What is more, it became clear that far from preventing the alienation of land from its traditional owners, '[m]ade into property, land becomes something to be considered independently of the social world' (Stead 2017: 80). In this process, and 'abstracted out from the intimate and particular relations of belonging, land is stripped of [much of] its agentive capacity and potential' (ibid.).

To understand the failure of existing legal instruments to capture the local people's relationship with their land, it is necessary to realise that, for many people in Papua New Guinea, land is not independent from the social world. Instead, in many local cosmologies, 'landscape and land are shared spaces of common imagination in which bodies and persons, institutions and histories are formed. The land is redolent with human presence, with the history of kinship, of people, of closely allied myth that inhibits space not as an overlay, but as its formation' (Leach 2014: 59).[6] At Frieda, the state used its legal apparatus to define and approach land as detachable and tradeable property, whilst the companies developing the Project over the years saw it as a commodifiable, loanable space, and a possible future site of an operating mine. Most Paiyamo, on the other hand, saw the land as neither a commodity nor a space but, to use a distinction proposed by Casey (1996: 26–27), as a historicised, relational place that was 'more of an event than a thing to be assimilated to known categories'. In local cosmology, land and places not only 'are', they 'happen', and 'it is because they happen that they lend themselves so well to narration, whether as history or story'. My hosts in Paupe were tied to their land by a complex set of localised relations embedded in their stories and histories that they were able to mobilise to sustain and develop what Leach (2014: 54) would describe as 'their fundamental conditions of life'. However, decades of engagement with the government and company representatives and the broader forces of the resource extraction industry meant that a process of commodification of land was already underway at Frieda, although not yet fully realised.

The lengthy process of planning for a mine at Frieda introduced the Paiyamo to different kinds of narratives about land and resources. It was symptomatic of wider processes unfolding across the country where, '[u]nder pressure from large-scale extractive industry, customary tenure [was] changing from the condition for constitutive and generative relations with land to a relation of property and control over land' (Leach 2014: 53). Over time, the Payiamo adapted the language of land ownership. But they did so largely insofar as property can be embedded within rather than detached from the world of social relations and become, as McCormack and Barclay (2013: 9) describe it 'a form of sociality that expresses relationships between persons and things and between persons with respect to things' that are mutable, not always clearly demarcated and subject to change. And even then it is difficult to call them landowners in the sense of owning the land. In Paupe, as in many other parts of the country, most people I met and talked to acted not as if they owned the land but as if they themselves 'belong[ed] to the land as much as it belong[ed] to them' (Sillitoe 1999: 348). However, a noticeable tension was emerging in the community as groups and individuals tried to reconcile local cosmologies of land with opportunities presented to them in the ideology of landownership. These tensions were particularly visible along generational lines in the community. While most members of the older generations continued to resist the process of commodification, turning instead to ancestral stories and *kastom*, younger communities saw within this process opportunities for individual wealth and access to infrastructure, services and development.

Inventing Landowners

In anthropological literature on the subject of mining in Papua New Guinea, we can read about processes of 'forging landowners' (Golub 2007)[7] and find the practice of landowner identification described as 'an exercise in the creation of legal fictions' (Jorgensen 2007a: 66). In a similar vein, writing about cultural responses to mining in Melanesia, Nick Bainton labelled landowners 'a fabrication of economic necessity and national legislation that simultaneously recognises customary land rights, but also demands a particular conceptual relationship, or separation from the land' (2010: 112). He went on to say that:

> many Papua New Guineans now find themselves responding to an "ideology of landownership" . . . that merges customary and legal definitions of ownership with expectations of development. However, the ground upon which they act out this ideology must also be productive, or possess some qualities likely to yield economic profit, or at least be subsumed within the broader area that a company, the government, or some outside agent is seeking to use. (Ibid.)

Bainton's remarks paint a picture of 'land ownership' as a conceptual category that places the landowners themselves but also the non-landowners, the company and the government in relation to the land as well as in relation to each other.

To that end, at Frieda, recognition of the area's rich mineral deposits turned the land into a particular kind of productive resource – one that bore a promise of widely conceived and much anticipated development. It may also be argued that the resource in question here is not the land itself. Rather, as proposed by Leach (2004: 42), the resource is 'the land involved in particular relationships'. It is through mobilisation of those relationships that the land becomes 'the site of real valuation [of] generative or productive relations between persons' and can bring about a desired effect (see also Hann 1998). In this process, landowner identification becomes an exercise in uncovering and recording relations between people more so than taking stock of relations between people and (their) lands – a highly politicised activity, playing out at the level of state legislation as well as local politics. As described by Jorgensen (1997: 609), 'all rights are subject to the provision that one is able to enforce them, and this has meant that effective land rights are ultimately based on membership in a political community capable of defending such claims against outsiders'.

State legislation actively promotes the idea that there exist in Papua New Guinea ancestral landowning clans that can be 'found' (Jorgensen 2007a). As a result, for many indigenous communities across the country, success in asserting a claim to benefits and/or access to agreement making at resource extraction projects becomes tied to their ability to successfully claim and demonstrate their relations with land and with each other in terms of a clan-based customary land ownership. But, as observed by Don Gardner (2004: 110) during his fieldwork among the Miyanmin communities living in the vicinity of the Frieda River Project, 'the processes of social life produce genealogically complex and heterogeneous communities' (see also Gardner 1987) that are a far cry from the structures imagined by the government and adopted by the companies. In his discussion on the peculiarity of group membership in the Frieda area, Gardner pointed to the high social mobility and dispersity that for generations characterised local group affiliations. He argued that the seemingly static clan names at Frieda were deceiving, as any single name covered a range of dynamic claims and relations, and every person could make multiple claims of belonging to different 'clans'.

Dan Jorgensen (2007a: 63) made similar observations among the Project's Telefomin communities and argued that in this part of the world 'it was always possible for individuals to claim affiliation along a diverse range of ties'. He described how, in response to the new types of relations and new kinds of descriptions brought into the area by the resource extraction interests, local people at Frieda 'invented clans – indeed, they invented several varieties of them – in a way calculated to match the expectations of the government and the mine developer, albeit in ways far removed from traditional ideas about the relation between land rights and collective identities' (ibid.: 66). Indeed, during my visits to the two Telefol villages up the river from Paupe, I often heard people say that although 'for the purpose of the Frieda River Mine' they belonged to one clan, they belonged to another one (or two) for pretty much everything else. For example, during one of my visits to Ok Isai, a young, educated man told me that 'from the history of the Telefomin, I am a Kiarikmin. But for the compensation over the mine – I am an Ountoumin'.

In Paupe, when describing the historical patterns of their social organisation, the Paiyamo spoke of lineage houses and their inhabitants. The longhouse communities emerged from these descriptions not as fixed social units but as, to borrow Jorgensen's term, flexible and elusive collective identities. The stories they told of house lineages described people brought together by mobilising a kaleidoscopic mosaic of relations, extending beyond blood and marriage ties and creating a dynamic and complex kin structure that was easily adaptable to changing conditions. The narratives described how, following the formulation of Paupe village they came to be known under a common appellation as the Paiyamo and how, in response to the company's demands for landowning, ancestral units, house lineages provided foundations for the emergence of landowning Paiyamo clans.

As I noted in chapter 2, those 'clans' were not demonstrative of a normative system of land ownership envisaged by the state and numerous business actors involved in the process of preparing for an operating mine at Frieda. Rather, they manifested relations between people who were 'mutually connected by knowledge and circumstances' (Strathern 2005: 68), bound by flexible links emergent from their shared interests in being together and in accessing the potential this being together can create. Also in chapter 2, I suggested that the idea of a landowning clan was neither trivialised nor externalised by the Paiyamo. It constituted an important factor, altering and shaping their world – awarding new significance to past events, reframing relations and creating novel opportunities for the future. The names and membership structures of individual clans changed over time in response to shifting contexts and changing local politics, and the ideas regarding the land's capacity to reshape networks of power and reconfigure relations were continuously reassessed in the context of the Project. For example, when Dan Jorgensen visited Paupe in 1984, the different groups he encountered called themselves Imaka, Koyali, Malipale, Pwimuku and Wasu. When I asked pastor Jacob about those names, he told me that Imaka and Koyali were now Melefale (Malipale), and Pwimuku and Wasu were now Suna. He went on to reiterate the names of the six Paiyamo clans: Imlou, Melefale, Nesiar, Ayamo, Inaku and Suna and told me that there are other names inside those groups but that those other names 'did not matter now', as they have since lost their power and importance.

Over time, clan land markings, initially serving an external purpose of asserting a stakeholder role and thus supporting claims to the Project's benefits, gained new internal significance. Following the 1996 land identification, I was told by the Paiyamo from all over Paupe that access to land became increasingly politicised and restricted. According to local narratives, the different Paiyamo lines always knew what land belonged to whom. Until recently, however, the land demarcations were rarely enforced within the community, and its members were free to travel freely around the area. But in light of debates about Project benefits, ownership rights, property and the need to fence the area off to protect it from an increasing number of people moving to Paupe and attempting to claim rights to the land, things were changing. During an evening conversation in Aston, an Ayamo man told me:

> Before, we all ran around the ground together. We knew the markings of the land, and which places belonged to whom, but there were no forbidden grounds. When we were looking for food, we would look for food together and it didn't matter whose land we were on. But now people think about the money. They stop others and say – you cannot enter. This is my land.

Another time, I recorded a Nesiar man say:

> Now the *kastom* of a white man is strong here and the power is coming out of the ground. You know – all the gold. And copper, too. Now things are changing, and you need permission from the *papagraun* to enter and use the land. Before this was not the case. Now you cannot just go whenever you want. They can use witchcraft against you. This is a big change – this matter of ground.

After a short pause he added: 'But this is a good change. If you open your lands, all the different kinds of people will come inside. They will find and take all the gold and you will be left with nothing.'

Bringing Back the Ountoumin

The expansion of the resource extraction industries brought with it new ways of eliciting, making visible and acting upon people's relationship with land. This led many groups to turn to ancestral stories for validation of their claims for recognition as *papagraun* – a Tok Pisin term traditionally used to describe individuals with a claim over particular areas of land, increasingly used in a communal sense to mean 'landowners' (see Filer 1997a and McGavin 2016). Philip Guddemi (1997: 641) described how, among the Sawiyanoo people in the East Sepik Province, 'the politics of land begins to have a different dimension as official ideologies begin to intervene in the ways land is used and thought about. Land becomes an aspect of a new dialogue, one that is partly about modernity and about a sense of proximity, or lack of it, to centres of colonial/national integration'. Mining projects tend to shift these scales even further. With their emphasis on land ownership, they link a range of people's aspirations and transformative capacities to given areas of land, and through the promise of economic benefits and development, they entail a sense of proximity to centres of not only national but global integration and bring with them a promise of participation literally in a world of opportunities. By their very nature, large-scale mining ventures redraw borders of influence and introduce new actors and novel factors into the local social, political and economic networks. In this process, land often gets disputed on economic as well as political grounds, especially when it is 'seen as facilitating one set of people's access to the world of development and money, instead of another's' (Leach 2004: 42). This is particularly challenging in a country like Papua New Guinea, where, as observed by Jorgensen (1997: 600), the 'characteristic landholding situation . . . is one where a range of overlapping rights may apply to any given territory, and numerous candidates may justifiably argue their claims in any given case'.

At Frieda, the complex history of land use and conquest, which included taking captives and kidnapping women, combined with the ways in which subsequent companies developing the Project went about identifying local landowners led to the legal recognition in 1996 of a re-emerged Ountoumin clan as the primary landowner in the Frieda area (see chapter 2). The court decision exacerbated tension and conflict, especially in the two Telefomin impact communities, where two other clans, Kiarikmin and Atempiakmin, zealously contested it. John Burton (2007: 175) described how, in many cases in Papua New Guinea, 'indigenous groups have been successful in pursuing their claims to land through various processes of landowner identification, including through litigation, but have found the outcomes extremely difficult to work with afterwards'. Frieda River is an example of a mineral exploration project where it was not just the indigenous groups who have found the supposedly successful outcomes of a landowner identification exercise difficult to work with.[8]

Xstrata found it hard to address and contain the discontent over Ountoumin claims to land and the 1996 court decision. Of key concern were the extensive political and often financial networks formed around and enabling the newly formed clan identity, which threatened to divert a large proportion of financial benefits from the Project to supposed clan members living often a fair distance away from Frieda, in areas not directly affected by the Project's activities. Several Community Affairs personnel I talked to at Frieda spoke about influencers from the Telefomin Station, Tabubil and further afield who provided traction and funding to the clan-based claims in exchange for a promise of participation in future mining benefits. They worried that, in the long term, unless some of those processes can be curbed, a large share of benefits will be diverted away from Frieda and communities directly impacted by the Project's activities – disadvantaging many groups and individuals and posing a risk of social unrest. To a significant extent, these worries were shared by many people in the impact communities, who were aware of the widening networks and growing external interests and claims and who could often be heard saying things like: 'We need to find a way to close these doors [to claims to land and benefits], they cannot stay open.' They were just unsure how to do this.

In an attempt to settle the matters, the company arranged a large-scale land mediation exercise in the hope that it might be able to convince people living in local communities to abandon the 'clan speak' and instead assert their claims to land ownership and therefore Project benefits on the basis of shared *kastom* and residency as, for example, the Telefomin communities of Wabia and Ok Isai. At the start of a mediation meeting I witnessed in Ok Isai in the last week of July 2012, a leader of the Community Affairs team spoke to the outcome of the 1996 landowner identification exercise and said: 'When the 1996 decision came, things got rearranged, clans appeared. Now, we have to go back to the true story and *kastom*. The politics were killing us over the past 16 years. It is now time that we resolve these political conflicts and move on.' Throughout the day, members of two other Telefomin 'clans' politically active in the community (Kiarikmin and Atempiakmin, or Kiariks and Atemps as they were more commonly called) questioned the legitimacy of Ountoumin and argued that all three clans should receive an equal share of royalties and compensation payments from the Project. As Bob, one of the local leaders and a member of the Atempiakmin clan, explained:

> Before 1996, there was never really any talk of Ountoumin. There were two Telefol descent lines here – Kiariks and Atemps. That was all. We came here from Telefolip and we settled here as colonists. But somebody somewhere came in and brainwashed some people into thinking differently . . . We all speak one language, and we would like equal footing – not that one will come [out] on top, above somehow. We all held bows and arrows and went to war here, and the war lasted a long time. We all got killed and eaten. We are one kind. I represent my interests as a Telefol and I speak for the Telefol. The Ountoumin are no more. Once a group is extinct, you cannot revive it. Once it does not have a language, songs, *kastom* and ancestral stories, you cannot bring it back. If I get adopted by a group and brought up by a group, I am a member of this group.[9] In the past, initiations played a crucial role here. Now we do not have initiations anymore. But three equally important factors continue to determine the membership of the group: biological relation (birth parents' groups); obligations of care and reciprocal exchanges; and living together on one land, working and eating and sleeping together. So here, we are three clans, but we are all Telefols, we have one language, *kastom*. We need to be treated as one community here.

When I talked to him later, Bob remarked that there were no 'clean blood' Telefols left in Ok Isai. 'There was a couple of pure blood fathers here, but they died last year. Through intermarriages, the individuals are no longer as clear as to what group they belong to as they used to be,' he added. However, the local leader of the Ountoumin, Fexian, dismissed those claims. He saw and used the court's 1996 decision as indisputable evidence of the clan's position and rights as primary landowners in the Frieda River Project area. He agreed, in principle, that all three Telefol clans in Ok Isai could receive an equal share of royalties from the future mine but strongly opposed what he saw as an attempt to introduce a new system based on shared land ownership. He continued:

> The Ountoumin were here before 1996. But in 1996, we came out clear to everybody. There has to be someone who really owns the land. All other mines have primary landowners, and this mine must, too. When it comes to receiving money, one clan has to come first. There can only be one, like there is only one driver in the car. They can then share the money with the others.

The next day, the Ok Isai land mediation was the key topic of conversation in Paupe. In Aston, pastor John explained:

> In the past, it was not a problem. Now is the time of money – and much untrue talk and false evidence keeps coming up. Before, we lived well. We had no issues when it came to land. Our ancestors marked our territories and we know the land well. But the Telefols they keep telling lies. They are just after the money.

'If you were to ask any Telefols they will tell you – Ountoumin is how they call a kind of arrow they use for shooting pig. The name belongs to an arrow, not to men. They all came out of Telefolip, they are all Kiariks,' Rhonda told me some time later, as we sat in her house sharing a plate of boiled root vegetables. Although pastor Jacob and some of the other local leaders occasionally claimed that if they were to follow the Telefols' logic and reasoning they too could claim ancestral links to the Ountoumin, the general feeling in Paupe was that the three Ok Isai clans should be treated equally. 'After all, really, they eat, live and sleep together; they are all one community. They are all Telefols,' they would say. Before the day drew to a close, I overheard a man in Aston tell the following story to a small group of people gathered in one of the wind houses:

> One time there was a man walking around the bush. He found a sago palm and saw an opportunity. He cut the tree down, stripped back its bark and extracted the sago pith. It was hard work and it took him a long time. But before he finished, the owner of the tree came by and saw what the man had done. The owner said to the man: I planted this tree, I looked after it; this is my sago. The man replied giving many reasons as to why the sago should belong to him: he has found the tree, cut the tree, extracted the

> pith and so on. The owner listened to the man's arguments and then replied simply: I planted this tree and so it is mine. What is the takeaway message of this story? The owner of the sago palm knows he is right and that the tree belongs to him. So he only has one story, one argument. That is all he needs. The man, on the other hand, knows that he did not plant the tree and that he was wrong to cut it down – for it was not his to cut. So he will be trying to work all sorts of talk and will come up with many stories to try to justify his claim.

The happenings I witnessed in Ok Isai during the land mediation exercise were illustrative of a process described by Guddemi (1997: 642), where 'the prospect of benefits and compensation to be paid out to customary landowners creates a scarcity value for land' and awards new interest and meaning to the 'subtleties of who goes first and who goes behind'. They also echoed Hirsch's (2001c: 307) claim that in mining contexts 'clans, like the future development of the exploration licenses, can only transpire by creating new forms of insiders and outsiders where these were not present before'– which in turn reflects the complex processes through which mining projects tend to redraw internal and external borders of power and influence and rearrange the grids of social and political relations. Melanesian societies are often justly attributed an ability to create and manage long and complex social networks, and lessons from the already operating mining projects in Papua New Guinea suggest that mining contexts provide means and incentives to both cut and expand them (see Strathern 1996), as local people 'seek novel ways to strategically manage . . . relationships in order to contain wealth within a limited sphere' (Bainton 2009: 19).[10]

Fencing Off the Paiyamo

In Paupe, the Paiyamo managed to reach an agreement on their clan structure and land boundaries, which they claimed were marked by ancestors but which, in their current form, both reflected and helped to shape some very contemporary relations and aspirations. Their recognition as landowners in the Lease for Mining Purposes area and an impact community meant that they stood to gain significant financial and non-financial benefits from the Frieda River Project. Over the years, the village became an attractive destination for people seeking to access those benefits and partake in the promise of development. 'We used to get excited about the new arrivals. But now we are worried,' Dinah once told me as we were walking the narrow path joining the SBU Camp and Aston. With that began a conversation, a version of which I had heard many times before, which described how for a long time the Paiyamo were intrigued by people visiting the area and the new opportunities these arrivals would bring. But over time, they came to be weary of newcomers. For example, they used to wait for the Sepik traders with anticipation, excited by the prospect of commodities and dry foods they carried up the Frieda River but were now worried about the alcohol and marihuana that were making their way to Paupe hidden at the bottom of the traders' long, motorised canoes. Their own experiences were further informed by the situation in Ok Isai and Wabia, where a large influx of traders and migrant alluvial gold workers became a source of numerous problems, often resulting in conflict and sometimes even violence (see chapter 1).

Not long after the land mediation meetings took place, a group of company and government officials visited Paupe to discuss creating what they called 'a social fence'. The aim, they said, was to protect the community from the influx of incomers that had already begun and which, the company representatives said, was likely to intensify dramatically once the Project progresses to the next stages of its life cycle. 'You are the gate to the mountains and to the project area' – the Paiyamo were told – 'Many people wanting to access the Project will come through here.' The proposed fence took the form of a community survey in which names of 'original and rightful' members of the community were to be recorded. Members of the company personnel explained to the Paiyamo that the aim of the exercise was to protect their rights from any false claims to land and benefits that other groups might make in the future. In conversations with me, they added that the exercise would also allow the company to limit its liabilities to recorded groups and people, thus making any future claims and obligations manageable. The Community Affairs officers were accompanied by two government liaison officials and an ex-kiap, all of whom were quick to say that it was up to the community to decide who is in and who is out. In preparation for the survey, the Paiyamo were asked to carefully consider who they were going to recognise as *papagraun* and who they are going to exclude. At the same time, they were told by both company and state representatives that in making that decision they should follow their *kastom*. 'If you do not follow your *kastom*, the community will suffer. Whatever ways you used to draw your lines before, use the same ways now. Do not come up with new ones. This will cause problems down the line. This work now – it is helping us and it is helping you. It is a big task and it needs to be done right,' said the ex-kiap.

The conversation that ensued inevitably led to a very contentious topic in Paupe – one of the rights of non-Paiyamo men married to Paiyamo women to settle in the village. Traditionally, upon the payment of the bride price, a man who married a Paiyamo woman was expected to move with her to his place of residence. The couple's children would not be considered Paiyamo, had no inheritance rights and could not make claims to land in the group's area. Many of my local hosts were concerned that their population numbers were low and talked about the importance of growing the numbers and strengthening the community's position, also in the context of exercising rights to land and, what follows, accessing benefits from the Project. Marriages within the community were actively encouraged. However, should a woman choose to marry outside, some local leaders were now prepared to accept both the husbands and the children as members of their clan. I use the term 'clan' deliberately here as other conditions had to be met – such as long co-residency, knowledge and following of the local *kastom* – for the husbands and children to be able to claim membership in the house lineage. There were a number of resulting cases in Paupe where husbands intentionally delayed the payment of the bride price. Not completing the customary transaction meant that they were not only allowed but expected to stay in Paupe with their wives and children until such time when they could meet their financial obligations towards their wives' families.

Over the course of the meeting, it became clear that views on the issue were split. While some clan leaders wanted to focus on following *kastom* and in doing so restrict access to benefits to their sons and grandsons, others wanted to grow in size, even at the cost of breaking with elements of local traditions. Pastor Jacob's view was very clear – 'if you are a woman and you are married somewhere else – you must follow our culture and you have to go – this is the way of our ancestors and the way of the place. Respecting this is more important than the money'. He was worried that if the community was to start doing things differently, even if this may be beneficial in the short term, it would create problems in the future. The fact that Dinah was married to a man from Amanap and lived in the SBU Camp with her family did not seem to affect his stance. But some of the others, including councillor Amos and pastor John, did not share that opinion. In their view, if a woman wanted to stay in Paupe and bring her children up as Paiyamo, it was her choice and that of her father's. 'If she wants to stay – who am I to say no? As long as she and her husband stay underneath me and he does not reach out his hand for the money first, it's OK. Myself, I want my daughters to marry here, have children and strengthen the clan,' said the councillor. The comment about reaching a hand out for money referred to a compensation distribution system introduced by the company. In that system, compensation money was split evenly between the households of Paupe. The money was distributed in cash, in brown envelopes handed over to male heads of households (with their wives witnessing) during public gatherings. It was down to household heads to distribute money to the members of the households. On one occasion, I overheard an exchange of strong words between Dinah and Peter, in which Dinah told Peter that, unlike him, she was not getting her benefits money in an envelope but was dependent on her father, Jacob, to share with her the content of his envelope and give her cash. The difference in views might be, to some extent at least, explained by the fact that while pastor Jacob's clan, Melefale, was the largest one in Paupe and held significant political power, the councillor's clan, Ayamo, consisted largely of himself, his two wives and their children, ten of whom were female.

In the end, a consensus was reached that it would be up to the leaders of individual clans to decide whom they would like to include and exclude from the company survey and, what follows, from official recognition as a member of the clan and a landowner in the Frieda River Project area. When the day of the survey came, company personnel set up a workstation in a wind house in Aston. One by one, members of Paupe's twenty-six households approached the officials to declare their names and clan affiliations. The resulting Community Register of Landowners included names of 201 people, 196 of whom identified themselves as members of the Paiyamo language group. While 141 were clearly identified as belonging to one of the five clans in Paupe, the remaining 60 entries either did not include information on affiliation to a particular 'subgroup' (clan), or claimed an affiliation to a clan under a name other than either Melefale, Ayamo, Suna, Nesiar or Inaku. When I asked pastor John about the new names, he raised his shoulders, shook his head and said:

> Money came up and suddenly plenty of people claim to be the owners of this land. What will happen next, will people start fighting over the ground? We do not know. Only God knows this. But many new things, names and claims will come up here because of the ground. In the future, people will be saying things. But those things you cannot believe.

The household survey was completed almost exactly a year after the Paiyamo presented The Book to the company. For my local hosts, however, the two documents were very different. The survey recorded *information* about individuals and Paupe's population in order to create a social fence for the future and to help the company as well as the community manage its networks and liabilities in the context of anticipated further claims to land and benefits. It had the potential to be important but bore no real significance and was awarded very little consequence at the time. The Book, on the other hand, communicated *knowledge* about people and was intrinsically tied to and demonstrative of the Paiyamo *knowledge* of the company. Its effect was largely unknown, expected to reveal itself at some point in the future, but its collaborative nature and novel format already awarded it significance. The Paiyamo used The Book to explain to the company (and, to an extent at least, themselves) their clan structure and relations between the different clans but also between the people, the land, the *masalai* and the ancestors. But possibly most importantly, they used it to affirm a relationship between them and the company, to signal their capacity to make their knowledge visible to the company along with a willingness to 'become stronger' for the Project's state and company stakeholders. They did so in the hope that this knowledge and these relations, presented in a specific form, would make them successful in influencing the agreement making processes and would eventually lead to life-changing financial benefits.

One evening, as we sat in Dinah's dimly lit kitchen discussing The Book whilst finishing a meal of cooked rice and tinned fish, Pastor Jacob told me that 'once it is written on paper, it is all there, and the government and the company will know'. Then, lowering his voice to a tone usually reserved for secret and sacred stories, he said that once the land boundaries are written down, the maps are drawn, and family listings are prepared, the company will know who owns which land, and the Community Affairs personnel will know how to direct the money.

The Mark of Good Leadership

Above, I described knowledge as information capable of causing effect in people and places and argued that it requires a knower to make it effective, linking the effect of knowledge to the quality of social relations and the knower's ability to make the knowledge (and the relations) visible in a particular form. A capacity for forming social relations, an ability to make those relations visible and the aptitude for mobilising them as to create new opportunities were some of the most valued qualities in Paupe and key attributes of local leaders, who owed their position within the community largely to their extensive knowledge and the ability to mobilise a range of networks and relations. These traits were precisely what made pastor Jacob a particularly prominent figure among the Paiyamo. If, as Slotta (2014: 634) tells us, '[e]stablishing personal authority goes hand in hand with establishing the authority of the sources of one's knowledge' then Jacob's ability to trace his knowledge back to the *haus tambaran* and the Bible gave him significant credibility. And if we accept, as suggested by LiPuma (1998: 67), that people tend to be evaluated on the basis of 'their personal biographies of activation of relationships', then his demonstrated and long-standing ability to engage with members of the community, as well as other local, company and state representatives at the Frieda River Project, evidenced his strong ability to mobilise and maintain relations.

Leadership positions among the Paiyamo have never been hereditary. Although being a son or a grandson of a distinguished figure in the community might have provided some men with an understandable and expected advantage, ultimately leadership positions were achieved on the basis of personal qualities, skills and achievements – one became a leader, a big man, through his actions and demonstrated capacities not through an official appointment. The past leaders, like Aiyapanai, were the war chiefs and the most eminent keepers of the knowledge of *haus tambaran*. They were also accomplished warriors and hunters, skilled in building houses, fishing, and growing taro gardens, able to look after their families and draw from their extensive knowledge and relations to explain unusual happenings and influence events. Their key responsibility was to ensure the safety and wellbeing of the community. In the cult houses, leadership structures were tied to a restrictive system of access to religious knowledge, as senior cult members held responsibility for carefully concealing and skilfully revealing sacred knowledge through ritualistic moments of revelation. This knowledge was emplaced in local social landscapes, histories and relations and served as a source of people's power and identity.

With the arrival of Christianity in the 1970s, religious knowledge, traditionally the domain of a few selected men made powerful by the culture of secrecy, was made openly available, with all literate people able to use the Bible to access 'the word of God' for themselves. As a result, 'the sociology of religious transmission has . . . been dramatically altered' (Lohmann 2001: 94) and gender and age hierarchy embedded in old leadership structures was challenged (see, e.g., Lohmann 2000; Brutti 2001; and MacDonald 2013). The new type of religious knowledge was not tied to particular localised places and provided people with a powerful sense of global citizenry and participation in a transnational and transracial system (see chapter 3). The coming of 'the white people' introduced additional novel political dynamics and further challenged local ideas of leadership. By appointing *tultuls* and *luluais*, the colonial administration introduced an alien structure of governance where expectations of leadership positions were demarcated outside of the local contexts and tailored towards externally rather than internally defined needs and processes. Following independence, the government of Papua New Guinea further reinforced those mechanisms by creating positions of elected village councillors and providing them with an externally entrenched authority and the power of a local-level government. In a parallel process, subsequent mining companies conducting exploration work at Frieda brought with them their own ideas about leadership, representation and – crucially – knowledge and tasked appointed individuals with representing communal interests on a number of often complex and contested fora.

When I asked the Paiyamo about the key attributes of contemporary leadership, they responded without much hesitation. They told me that the 'leaders of today' had to be smart, 'good in the head' and able to read and write. They ought to be confident, not afraid to speak out and know how to talk to the white people and the company. Ideally, they should be able to speak English or at least hold a very good command of spoken and written

Tok Pisin. It was important that they knew the local *kastom* and land, but it helped if they were also able to 'speak the language of the company and the government'.[11] As it was explained to me in Paupe: 'We all speak Tok Pisin here now. But our Tok Pisin is different to the Tok Pisin of the company and the government. And the Tok Pisin of the consultants who come here is of yet another kind.' In the Project's impact communities along the Frieda River, Paupe included, the Project's demands for a new kind of knowledge and the newly emergent means of achieving leadership status often led to tensions between older men, keepers of the *kastom* and knowledge of *haus tambaran*, and younger ones, some of whom had formal education and had ran small-scale local businesses, gaining experience in business networks, transactions and negotiations. The former strived to assert the ongoing importance of *kastom* and traditional knowledge in the processes surrounding planning for a mine at Frieda. They lamented the fact that the latter were increasingly disinterested in ancestral stories and sacred knowledge, preferring instead to focus on formal education. At the same time, they loudly proclaimed the relevance and leverage of school education and technical training in the context of the Frieda River Project. All communities took great pride in their school graduates and saw formal education as a way of countering some of the power inequalities playing out in the Project, in which the company and the state's preference for particular kinds of knowledge, and stakeholders' uneven ability to engage with it, played a large part. Many of my interlocutors remarked that the leaders for the future should be fluent both in traditional and 'school knowledge' and able to draw power and leverage from a wide range of past and contemporary relations.

The Future Behind

Although local ideas about leadership and power attributes evolved over time in response to shifting conditions and changing expectations, they remained intrinsically tied to people's knowledge and relations and their ability to act on those and make them effective. Strathern (1991: 199) proposed that 'if social life consists in making the internal capacities of persons visible, then it rests on techniques of revelation which must also continually externalise and thus re-present these capacities in new forms'. Extending this observation to knowledge, my argument is that, like relations, the effect of knowledge is also tied to the knower's ability to make it visible to his or her audience in a particular form, and at a particular time. In Papua New Guinea, mining interests introduced new ways of 'doing knowledge' and getting to know local people. Labels such as landowners and impact communities played an important part in this process (see chapter 2). These classifications have been recognised by Hirsch (2002: 370) as 'the sort of systematic, general, and accurate knowledge that the bureaucratic form of nation-states, including other modern bureaucracies, need for their operation, and that the modern social sciences act to supply. Such knowledge is sought because of the aspiration to formulate explicit and general truths about people'. At Frieda, in the context of the Project, the Paiyamo found themselves faced with a difficult task of demonstrating their relations, knowledge and capacities to the company and government. They did so by becoming known to them through interpersonal engagements and relations. But, in the long run, this tactic failed to achieve the desired effect. The mine stayed stubbornly hidden, the company kept demanding further information and, with a newly announced extension to the feasibility study, the promised development remained a distant prospect. In order to, in their own words, 'make themselves stronger for the company and the government', my hosts in Paupe embarked on a process of making their knowledge visible in a form resembling company documentation and the ways the company and the government 'did knowledge'. In doing so, they were driven by a belief that their knowledge, presented this way, would achieve a desired effect.[12] The Book is a result of this process. It presents knowledge about the Paiyamo in terms of clan listings and GIS maps of their lands and uses the story of Melesinapi to demonstrate the Paiyamo knowledge of and their relationship with land in the project area, and with each other. To understand what The Book is and what it does, it needs to be read as a product and demonstration of a set of expectations that the Paiyamo hold about these relations in the wider context of the Frieda River Project.

Throughout my fieldwork, topics of relations and transformative effects of knowledge featured prominently in the stories my hosts told me about the Frieda River Project. The deposits were hidden in the rocky grounds of Frieda by Afek, the great ancestress of the Telefomin, and have been guarded by a *masalai* since. There were people at Frieda who knew his hidden name and, by revealing this knowledge, could enact a relationship with the *masalai* and ask the spirit to stop hiding the minerals. I was told on numerous occasions that people did not understand the true value and meaning of the deposit until the first companies arrived in the area, drawn in by its generous mining prospects. The companies brought with them new kinds of knowledge and revealed the land at Frieda as a new kind of resource with a novel capacity to transform the lives of the people, expand their relational networks and bring them political recognition and development. As the years went by, more people and companies made the long journey up the Sepik and on to Frieda. Not only did they bring with them visions of new opportunities for the future, they allowed the Paiyamo to reinterpret old stories and reveal new meanings in events that had already taken place, like when the Paiyamo say that they did not realise the importance of their 1942 encounter with the German woman until almost thirty years later when the first CEC representatives arrived on the banks of the Frieda River. My hosts in Paupe were very mindful of the fact that the full meaning of events could never really be known at the time of happening but was expected to reveal itself over time. The effect of revelation played a crucial part in the rituals of the *haus tambaran* and continued to constitute an important practice through which relations were forged and knowledge awarded meaning and efficacy, and through which 'social life [was] experienced and conceptualised' (Slotta 2014: 627) at Frieda. The Paiyamo used the Tok Pisin term *bihain* (from the English behind) to speak of the future and explained to me that in many ways this future has already happened but it has not yet revealed itself.

The idea that events are not knowable as they happen was discussed by Simon Harrison (2001) in his reflections on a return trip he made to the Sepik village of Avatip twenty years after he completed his initial period of fieldwork there in the 1970s. Having revisited old relations, and having been reminded of past events, Harrison claimed that to the Avatip people most events did not bear significance at the time of happening. Their importance, he argued, became apparent only from the perspective of the time past. He wrote:

> It is as though some acts have no determinate meaning at the time they happen, but rather a range of potential meanings, because it is only the subsequent course of events that will end this indeterminacy, clarify and define these acts and make them appear to have a proper place in the past. Only in hindsight do they become completely visible, when subsequent history has turned them retroactively into fully perceptible events at last. (2001: 9)

Much of Harrison's observation is reminiscent of my own fieldwork in Paupe. With one exception. The Paiyamo did not view events as singular objects of knowledge that could, even from the perspective of time and under the right conditions, become completely visible. In this geographically distant corner of Papua New Guinea, any event was, to adopt Strathern's term, dividual – made up of multiple meanings, elements and relations – and contained within it transformative potential that could never really be *fully* known. What is more, this potential had no fixed measure or identity. It was continuously reshaped, animated by people's actions and relations through which the event itself became, to an extent at least, knowable.

To some extent, this might explain the relatively low-key character of the ceremony in which The Book was presented to the company officials. That morning, the document was described as important, a big deal by both pastor Jacob and the company representatives, but the event felt anything but – after all the lively build-up from the night before, it was brief and poorly attended. Comparing the two gatherings, it would seem that by the time the brown envelope containing a printed copy of the document exchanged hands, the real event had already taken place as the Paiyamo had strategically mobilised a range of relations to put The Book together. In a moment of unity, they awarded it with efficacy by endorsing it and making their knowledge and relations appear in a particular form. They hoped that the document would elicit new relations and lead to new recognition as well as greater power and status, and eventually enable their transformation into a financially well-off and 'developed' community. In this context, The Book was demonstrative of a process through which the Paiyamo embraced and utilised new relations and forms of making knowledge visible in order to gain access to the world beyond the horizon and to things they imagined as much greater than the ones made available to them by their current relations. They invested all these anticipated outcomes in the document and waited for its effect.

Conclusions

In any legal contract, let alone one concerning potentially billions of dollars, a complex corporate entity and a state that has not yet quite reached democratic maturity, it is difficult to have a party that changes entirely, constantly manoeuvring things and presenting them as if nothing has changed. It is even more difficult if these processes of change are not simply triggered by resistance or a desire to get on the economic register but if they constitute an inherent cultural property of that community. The kaleidoscopic nature of the Paiyamo means that the Paupe community stubbornly resists any form of definitive classification, even if their appearances (as presented in The Book, for example) might suggest otherwise. Note that The Book states quite explicitly that the Paiyamo reserve the right to change their structures at any point, in response to shifting interests and changing conditions. In a similar vein, the company's attempts to create a register of the members of different clans with the hope that by limiting the number of potential future claimants they will be able to limit their liabilities are likely to produce uncertain results. This is because documents may be successful in recording information, but – on the banks of the Frieda River at least – it is dynamic and entangled relations that 'do knowledge'. Removed from a complex set of relations that animate it, The Book makes little sense and loses all its transformative power. In order to understand what it is that the Paiyamo did with The Book, it is necessary to embed the document within a complex and contested process through which the company tried to get to know the people using particular kinds of descriptions, and a parallel process in which the Paiyamo used The Book to gain knowledge of the company by judging the effect of making themselves visible in a particular way.

Sillitoe (1999: 331) argued that 'land situates humans in the world and symbolizes continuity'. But this is not to say that it somehow denies or prohibits transformations. Quite the opposite. Seen as a product of dynamic relations, land and places 'witness and incorporate change' (Telban 2013: 86). To my hosts in Paupe, land has always been a resource, caught up in discourses of identity, politics, power and social mobility. It was never just a site where things had happened and continued to happen, but the land itself happened, animated by a range of relations that mobilised it, utilising its great capacity to transform networks and to award new importance to people, places and events. The mining companies revealed new meaning and value in the area's wealthy deposits of gold and copper and tied them to an array of future promises. The Paiyamo were not surprised that the novel set of opportunities should stem from their places and be born out of their relationships with land, the *masalai* and with each other. Making land an object of relations was what they have been doing all along. But it would be misleading not to acknowledge the ways in which changing concepts of ownership and communal and individual responsibilities developed around the Project, have challenged some of the long-standing local ideas and, in some cases, become an issue of concern in Paupe.

Notes

1. Spelled in the document as Mele Sinapi.
2. K13,000 amounted to approximately £4,000 in 2011.
3. See also Brutti (2007) for a description of how the Oksapmin people rediscovered river spirits and sacral sites to claim compensation for the ecological impact of mining operations in their area. In this context, 'the ground and, in broader sense, the earth find again their original importance [as] the sacred sites are re-interpreted and perceived as guardians of land rights, held by mythical actors in the beginning of time' (2007: 254).
4. See also Dwyer and Minnegal (2014), who use examples of locally produced maps to analyse the role of abstraction in social change on the fringes of the PNG LNG Project.
5. The use of ancestral songs is well documented across Papua New Guinea, also in the context of resource extraction projects. For example, see Gillespie (2005, 2013) and Bainton et al. (2011 and 2012) for a case study from the Lihir Island, where the lyrics of some songs were altered to incorporate changes in the physical landscape resulting from the impact of the Lihir Mine. See also Weiner on how 'geographic dimensions of human community' (2004: 10; 1991 and 2002)

were made visible through song as well as myth and poetry among the Foi for the Kutubu Oil Project; and Wardlow (2004) on how names of places were incorporated into Huli songs at the Mt Kare Project in Papua New Guinea's Enga Province.

6. For example, in his evocative account of narratives of land and value among near the Ramu NiCo mining project, James Leach (2014: 59) described the social world of the Reite people (Rai Coast) as 'one in which land and people are explicitly and consciously interwoven in processes of social formation, production, and reproduction. Kinship is rooted in particular places; land underwrites the social relationships it nurtures. Those relationships manifest in persons, and in things, making these creations aspects of the place itself, drawn from and feeding back into a specific emergent productivity'.
7. Golub (2007: 88) described the Porgera Land Study (aimed at identification of landowners in the Porgera mine area) as a 'process through which corporate entities like "clans" creatively emerged in response to the land study's elicitation of them. We have seen that the land study was indeed "forged" – it is neither a direct translation of a pristine, timeless Ipili social structure, nor a travesty in which Ipili culture was misapprehended. The land study in Porgera was instead a creative augmentation of Ipili social organization according to government requirements and Ipili predilection'.
8. Such comments are indicative of the increasingly recognised 'risks attached to current processes of development where intervention is framed in terms of assumed cultural entity rather than in terms of malleable and dynamic sociality' (Gilberthorpe 2013a: 263). They have been linked to a practice common in the extractive industries that Emma Gilberthorpe glossed as *culturisation* and defined as an imposed 'ideology of *culture-ness* that connects people to place in pursuit of the products of development – infrastructure and services, private property, individually accumulated wealth, and independent investment' (ibid.). To that end, she warned, 'the industrial encounter can be divisive to the complex systems of sociality that help maintain economic and social security for all' (ibid.: 273; see also Gilberthorpe and Sillitoe 2009 and Gilberthorpe and Banks 2012).
9. Here Bob made reference to the story presented in chapter 2, which speaks of a pregnant Ountoumin woman being kidnapped and held captive by the Telefols during the period of tribal fighting. The Ountoumin use the story of the woman and her son to claim distinctive identity as the Ountoumin clan.
10. For example, Kirsch (2001, 2004) argued that, on the island of Lihir, claims of property and ownership shortened social networks and served to limit the company's liabilities, while claims of compensation brought them into view, expanding the possibilities for participation and implicating the company in extensive chains of events. Also writing about Lihir, Bainton (2010: 129) described how,

in the context of mining benefits, 'previous beliefs about an expansive and inclusive network have been reconsidered, if not altogether dissolved' and local big men who traditionally depended on expanding multiple layers of their networks now preferred to keep them out of view in order to manage their networks in the context of the mine and to 'deny the connections that can provide others with access to benefits' (2009: 18).

11. The new demands opened the doors for external influencers who used their personal attributes to try to gain positions of power within the communities. Those people tended to assert a form of historical or interpersonal connection to claim their right to reside in one of the impact villages. They would then use their personal qualities (e.g. self-confidence) and skills (e.g. formal education, command of English) to claim the position of community representative, revealing to the community their knowledge of international institutions and, as described by James Slotta in the context of the Yopno-Uruwa-Som Conservation Area, 'presenting themselves as the necessary mediators between an "out-of-touch" community and a knowledgeable, powerful, and yet obscure world of transnational actors' (2014: 626). More often than not, such influencers acted on their individual interests more so than on the interests of the community and served as gatekeepers to stop other community members from engaging with the company and government representatives. The company managing the project was aware of this but found it difficult to get past them and limit their voice in meetings and negotiations.
12. Using an example of a petition prepared by the West Ningerum Pressure Association, Crook (2007a) illustrated a similar process of strategic presentation of knowledge at the Ok Tedi Mine. Placing his analysis in the regional context of knowledge practices characterised by concern with revelation and concealment, he described how the authors of the document chose to present the evidence of land degradation in a language comprehensible to Western science, and excluding customary explanations.

Illustration 5.1. A flag pole in Aston displaying the PNG flag. Photograph by the author.

5

Precipitating the Mine

Introduction

At one point, I asked a group of children in Paupe to each draw me two pictures of their village. I asked that the first picture presents Paupe as it was at the time. For the second picture, I asked them to imagine what life in Paupe will be like in the future. The results of this exercise were quite striking, although not surprising, pointing to a particular imagery inherently linked to the prospect of large-scale mining, money and 'development'. While the first set of drawings tended to present houses built on raised platforms and made of bush materials, taro plants, fish, banana trees and, in few cases, sketches of pigs and helicopters, the second set was markedly different. Images of helicopters and diggers dominated children's drawings representing the future, with many featuring roads, buses, bicycles, money, ice cream cones, bottled water and houses that were, a number of them explained to me, built with timber and covered with copper clad roofs and therefore better than the current buildings in the village (see Figures 5.1. and 5.2.).

The results of this simple exercise were indicative of the anticipation with which the Paiyamo waited for an operating mine at Frieda which, they hoped, would bring about the much desired and eagerly anticipated development. For them, the mine bore a promise of better lives to be lived on the banks of the Frieda River and elicited a powerful and attractive imagery of a bright, affluent future that could, one day, become their reality.[1] They longed for access to material wealth and a world of opportunities they now knew existed but felt they were being denied, and they believed that the mine would constitute a portal through which they would gain access to this world, which they saw as superior in many ways to their own. This chapter presents local narratives of desire and expectations set against a backdrop of the effects of the mine at Frieda, which have already begun to show. It

Figure 5.1. Drawing of Paupe in the present. Photograph by the author.

discusses how, in the government's notable absence from the area, and deficiencies in services and infrastructure delivery, the Paiyamo looked up to the company for what traditionally would have been considered responsibilities of the state. Continuing to explore the interplay between prescriptions of stakeholder roles and their enactments in the Project location, and to further highlight the difference that perspective makes, the chapter discusses expe-

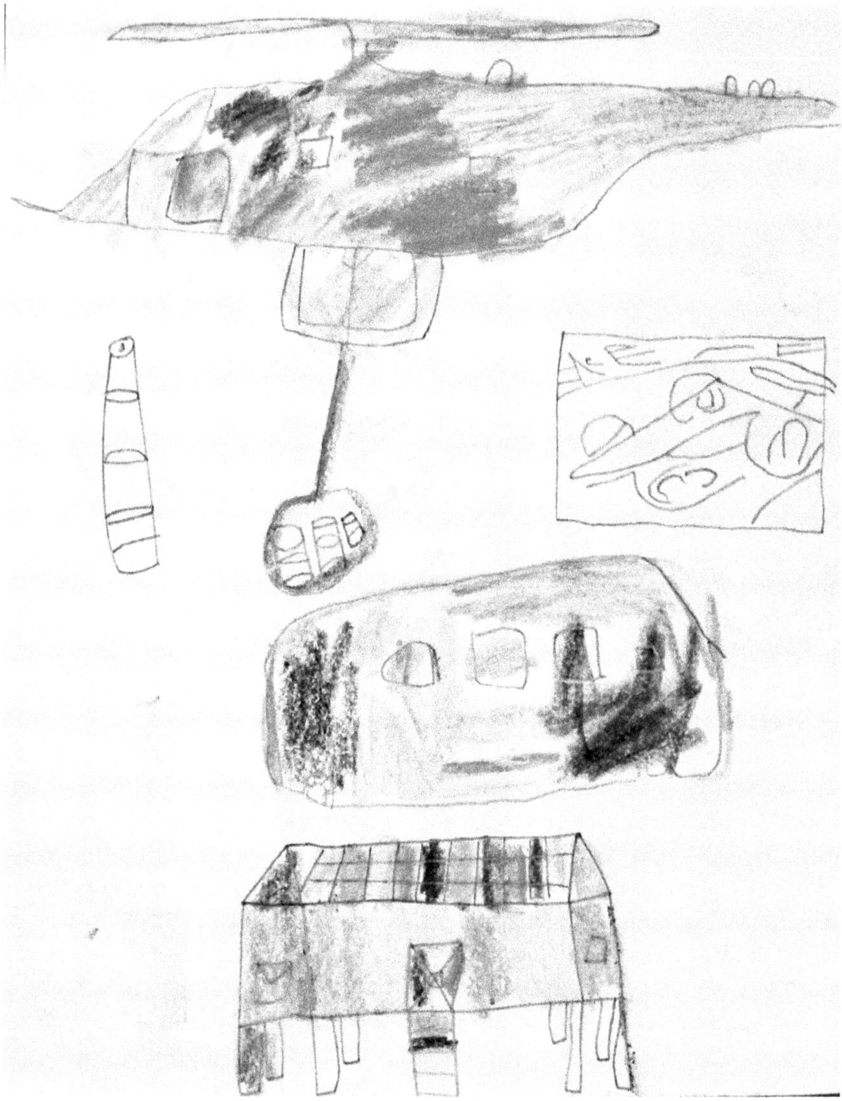

Figure 5.2. Drawing of Paupe in the future. Photograph by the author.

riences of Community Affairs personnel at the Project, who were entangled in dynamic relations they developed with the Project's local stakeholders. It uses ethnographic material collected among the company personnel to describe how members of the site-level Community Affairs team were left to bridge the relational gap between the company and the increasingly impatient and disillusioned local communities. The chapter shows how the team's professional roles were shaped by the discourses of corporate social responsibility and sustainable development and follows them in describing ways in which the company interpreted the latter in a way that allowed it to limit the provision of community development assistance in the area. It argues that this was made possible by the way in which the company differentiated between a resource exploration project and an operating mine – demonstrating the importance and value of an enquiry into multiple temporalities and constitutions of 'a mine' at Frieda River.

The Effect of the Mine

Early on in this book, I presented a story of Afek, the great ancestress of the Telefomin. I followed the Paiyamo in describing how, before she left, Afek hid a stone called *bangri* in this hilly and lusciously green part of Papua New Guinea. She asked *masalai* to hide it and to keep it hidden until the time was right for *bangri* to be revealed. Although my hosts in Paupe told me that this particular story belonged to the Telefols, they said that it was their story, too, and used it to explain the origins of the mineral deposits in the area and to describe the beginnings of the mine at Frieda River. In the introduction, I described one particular occasion when, after telling the myth of Afek and *bangri*, the teller went on to say that 'it is an ancestral story but now things are happening and the story is becoming true. It is an old story but it is true now' and further that 'the spirit was hiding the stone from many companies. This is why they did not extract the gold and copper here. But it does not hide it from Xstrata. Before we did not want the mine, but now we know how important development is, so we asked the spirits to stop hiding the stone'. This last comment offers a valuable and intriguing insight into the ways in which local communities came to understand the Frieda River Project in terms of knowledge and relations through which aspects of their own world and their own lives were revealed to them and to others, creating a range of new opportunities for the future. It is embedded firmly within local epistemology and emplaced ways of world-making and brings together a number of themes that have been explored in the previous chapters of this book, such as uncovering the value and meaning of ancestral stories – the aesthetic of revelation – and ideas of land and its resources and, crucially, local ideas of personhoods and social relations. It also points to the ways in which the Project's local stakeholders made claims about their knowledge and relations in an attempt to respond to the serious power inequalities at play and assert the validity and efficacy of their life-world in the context of the Project. The ethnographic material presented in this book comes together to create an interpretative framework in which the story of *bangri* and the origins of a mine at Frieda are made powerful by an enacted belief in its efficacy. Told locally, the story is made true by the effect it has already started to have on people and places.

When I first arrived at Frieda, the Project, under Xstrata's management, was nearing the end of a full feasibility study. The levels of anticipation in Paupe were high, and the Paiyamo hoped that, under the company's management and as an effect of the successful performance of stakeholder roles and relations, the mine at Frieda would finally reveal itself, bringing with it new opportunities and the long promised and much awaited development. It is difficult to pin down with all certainty the beginning of development discourse at Frieda, although many of my local informants would trace its history back to the early 1960s, when the first government patrol moored its boat on the steep banks of the Frieda River. Even fifty years after the event, the latter was conceptualised largely in terms of opening up new pathways to modernity and 'cargo' and strongly associated with a new and intriguing imagery of great opportunities for the future. The kiap brought the Paiyamo together in the village of Paupe and told them stories of Papua New Guinea, linking them to more people than they had ever encountered in their lifetimes. The idea of a nation, even pre-independence, exposed the Paiyamo to networks more extensive and more inexhaustible than their already complex political, cultural and economic systems. It made them, to borrow a phrase from Benedict Anderson (1991: 6), members of an imagined community made up of people who 'will never know most of their fellow-members, meet them, or even hear of them, yet in the minds of each lives the image of their communion'. This imagery of new communities, networks and relations was expanded further when, in the early 1970s, the first company began conducting on-site exploration of Frieda's rich gold and copper deposits and when, only a few years later, a Baptist religious revival revealed the Paiyamo as the children of God and members of a global, Christian community. As in many other places across the region, the mutual implication of mining and Christianity in local perception granted people 'membership in a world transcending local horizons – a world in which local people were always already natives of a distant land' (Jorgensen 2014: 24; 2005) – and became fundamental to the process through which the communities engaged with and claimed participation in the project of development and modernity.[2]

Initially, at least, the Paiyamo drew their theory of development from their encounters with the missionaries, company and government representatives and the stories those new arrivals told of the lives lived by people in distant lands, beyond the horizon. These stories offered the Paiyamo glimpses into what their own lives could look like, and the nexus of relations developing around industry's growing interests in Frieda's gold and copper deposit provided them with the means to enact a belief in 'development' and 'the mine' and to make those new lives happen. In their own way, the

Paiyamo embraced a version of the Euro-American ethos of development as 'progress' and, in a characteristically open and flexible manner, were intrigued by a version of themselves that they read from watching the state and the company. Over time, the effects of this enacted belief were gradually revealed to the Paiyamo, allowing them glimpses into the transformative potential of the mineral deposits and the efficacy of the juxtaposition of social relations that mobilised it.

'Before our lives were rubbish. Now the company is here and we are OK,' they could often be heard remarking. The compensation payments and employment opportunities offered by the company supplemented occasional income generated through intermittent alluvial gold mining activities and allowed them to take part in burgeoning commercial markets and the cash economy. The Paiyamo greatly enjoyed access to a range of foods and consumables available for sale in Paupe village shops or on the Ok Isai waterfront and took every opportunity to travel further afield, to Vanimo, Mt Hagen or Wewak, which informed their ideas of development and modernity. Utilising personal relations with the company personnel, they turned cash into commodities brought in on the corporate fixed-wing charter from Port Moresby and Australia, and were fascinated by the efficacy of money and the ways in which those relations granted them access to consumables they associated with development. This is not uncommon in Papua New Guinea. 'The desire to establish social ties with those who will then bring them the material wealth and services reflects a distinctly Melanesian approach to the process of change and the acquisition of goods' (Filer and MacIntyre 2006: 225) as people hope that forming social relations with employees of the mining company 'will provide access to goods and services and help to bridge spatial, social and ideological divisions' (West 2006: 310). The Paiyamo eagerly used every opportunity to cross the tall gate at the end of the muddy path to enter the Airstrip Camp, where they worked on developing relationships with the company employees and where they carefully observed infrastructure, goods and services available to the personnel and imagined that this could one day be their lives, their future. 'Airstrip is a sweet/happy place,' Dinah would often remark, seeing as men, even those not employed by the company, used every excuse to go to the airstrip and often spent long hours sitting outside the old Community Affairs office building, chewing betelnut, smoking, telling stories and exchanging information.

When talking about an anticipated mine at Frieda, people in Paupe – men, women and children alike – spoke with excitement about what they came to see as an attribute of development and developed communities. When asked, they could often be heard saying that when the mine comes they will live in good houses and eat a variety of store-bought foods. They will have money and will be able to afford clothes, sweets, cosmetics, white goods and electronic devices. They eagerly anticipated the mine access road, which, they expected, would connect them with Vanimo and maybe even Tabubil, and awaited other services such as schools and hospitals, which would be built on the banks of the Frieda River. Some people hoped that the mine would start operating public charter flights to places like Mt Hagen, Wewak and Lae. Others, especially young men, talked about business and employment opportunities that would open up when the mine starts operating, as happened in Madang. 'People there used to be like us. But now they have roads, good buildings and businesses. And it is all thanks to Ramu NiCo and mining. They now have a mine and they are alright,' they told me, referring to the nickel mining project in Madang Province. Crucially, many of my interlocutors in Paupe believed that, with development, they would gain skills, power and influence and, in many ways, become more equal to the company and state stakeholders involved in the Frieda River Project. Preparation and presentation of The Book of Paiyamo was part of this process and a way in which the Paiyamo wanted to demonstrate their stakeholder (landowner) form and capacities with regards to the Project.

When describing their new, better future, they would repeatedly use an analogy to 'the ways of the white people', which they saw as symptomatic of development and as contrasting sharply with aspects of their own contemporary lives. They often talked about good lifestyles enjoyed by the white people, an impression that was heavily informed by their encounters with

Illustration 5.2. 'We watch movies, we go into town, we know how it is . . .' Photograph by the author.

ex-patriate company workers and contractors, as well as American movies imported from Indonesia on pirated DVDs and watched on rare movie evenings in the village. On those occasions, a generator would be switched on and members of the community would gather around an old, large TV set to watch romantic comedies and action films. 'We watch movies of the white men and they all play, they are all happy. And this is what we want here, too,' one of the local leaders said to me as we were watching a movie I did not recognise. There were children in Paupe named after actors or movie characters (e.g. Sylvester, after Sylvester Stallone), and my hosts often made references to stories told in the movies to explain to themselves the ways of the white people and development that will, one day, arrive at Frieda. A young Paiyamo man told me, referring to the anticipated royalties and land compensation payments:

> We watch movies, we go into town, we know how it is. And this kind of life will come here, too. Every fortnight, *papagraun* will come to get money, get benefits. It will be like this. On payday there will be races – everyone will queue outside the bank until they get tired. Workers will get their fortnight and all *papagraun* will get free money.

Among all the excitement elicited by the promise of development and the new, prosperous future, the Paiyamo voiced their worries related to the negative effects of the mine. For many of my hosts, an operating mine at Frieda bore with it a promise of wealth and opportunities. At the same time, it posed a real threat to the local social and environmental systems. These were strongly informed by their experience of the impact of alluvial gold mining in Ok Isai and Wabia, as well as their knowledge of other mining projects in Papua New Guinea and the Indonesian province of West Papua. 'I have never seen the Fly River, but I saw the photos. So for me, I know, mine is not a good thing. It brings good services, but it destroys our environment,' one of the men in Paupe explained to me. 'I too heard that when the mine comes Frieda will no longer have fish. The pigs will be OK. But when the water gets polluted it will turn yellow or green and the fish will die. Not in the small creeks, though, and gardens and bush will keep providing food. If we forget how to fish, hunt, grow food in gardens, if this happens our lives will be horrible. We will be living on money only. But now we are rich in everything,' I heard from another.

Most adults in Paupe demonstrated great curiosity about the mine, development and what this new world will bring. But they also spoke openly about their concerns related to the future of their *kastom* and the world of the ancestors, which, they felt, was slowly beginning to disappear. During one of our conversations, councillor Amos stopped me half way through asking a question and said:

> This company is good, and they are doing important things here. It is us who need to change our ways. The money comes in and what do we do? Men buy beer. They don't want to look after their families. Our role as fathers is to make sure that we bring the children up in a right way. Jacob followed his father when it comes to knowledge. Aiyapanai taught him everything. But someone has to follow Jacob, too. When this mine comes, the difficulty and need for knowledge and leadership will not be fake. The struggle will be real.

A few weeks later, over dinner in Dinah's kitchen, pastor Jacob paused after telling us an ancestral myth of a snake from the Arimoku Mountain – the second most powerful *masalai* of the Melefale after Melesinapi. He took a slow sip of his coffee and said:

> When the mine starts, white men will come, our *kastom* will end. No-one will be building traditional houses and people will forget how to grow taro and wash out sago pith. In Tabubil, next to the Ok Tedi mine, they now have a *kastom* house, but it is a museum, a pretend-house only, something to show the kids. Many people don't have it easy there, they live with worry. They already forgot how to work gardens, and their sago palms have dried up. Here too, some kids no longer know our language, songs and stories.

A short while later, he turned to Mawi, who was two years old at the time and clearly getting bored of the conversation. He sat her on his lap with her face towards him and, holding both her hands, playfully told her that when mining starts, they will be eating rice, tinned fish and crackers *every day*. She smiled and clapped loudly, clearly liking the idea.

Like Jacob, Dinah too had mixed feelings about the mine. She told me on several occasions that, like pretty much all people in Paupe, she wanted to see development at Frieda. 'Many people have never seen a car or a road, or a ship. Others have been to towns and have seen it. Those who haven't, want to. Those who have, want these things here,' she remarked. At the same time, she worried that the new life would benefit some people but not others, adding to the politics of inclusion and exclusion that have already emerged in the context of defining and enacting the landowner stakeholder category. Her concern was that all hardworking people from Paupe will go to work for the company and that those who stay behind will not want to work at the mine site but will not want to put effort into maintaining houses and gardens, either. She feared they would put pressure on those who are working to share their money with them. 'So the hard working people will all have to work even harder, and the others will just stay in the house, do nothing and wait to be fed,' she lamented, whilst walking up a steep track leading from the SBU Camp to her banana garden further up the hill.

Illustration 5.3. Dinah scaling fish during a successful fishing trip. Photograph by the author.

I once invited Dinah and the other women from Paupe to participate in a focus group I had organised at the SBU Camp's church. It took place on a Wednesday morning, immediately after the weekly women's worship. Most of the service participants stayed behind to talk to me, sat in a semi-circle on the church's uneven and cold concrete floor. In response to my initial question about their attitudes towards the mine, they said that development [*sic*] was a good thing and that they were anxiously waiting for the mine to start operations. When I asked about the future impact of the mine, they told me that they did not know what would happen. 'We have no experience of a mine. This is a first time for us. Will we live well, will we eat and work well – this we do not know,' said one of the women, Jenny. 'We live deep inside the bush, and we do not have development here. So we need it to come and then we will see and we will know,' added Tela, in a comment that reflected knowledge through hearing as validated by knowledge through seeing (e.g. when the story of Semiti and Mt Usimi was accompanied in the *haus tambaran* by the showing of a piece of crystal stone to evidence that the story was true).[3] The women talked about their worries associated with immigration and their husbands marrying again and bringing their new wives to Paupe. Male polygamy was relatively common in the community. It was linked to the wealth and resourcefulness of men, and the women expected cases of polygamy to increase once the mine starts operating and the anticipated substantial financial benefits reach the community. They feared that new people who arrive at Frieda, either through marriage or in search or employment opportunities, will bring with them problems and illnesses that 'belong to different places'. As such, the Paiyamo will not know how to deal with them and will be overwhelmed or harmed in one way or another.

Mostly, however, the women expressed their concerns about the impact the mine's operations will have on their household structures and associated daily tasks. 'When the mine comes and all the men go to work, it will be just us and children in the village and it will be hard. Who will help us clear the land for gardens, cut big trees, hunt for meat, build our houses and fences? You need both men and women in a village, then everything can get done and all is well,' a number of women exclaimed, interrupting and finishing each other's sentences. When Dinah suggested that women too will be able to go to work, find employment as cooks or, like Rhonda, laundry assistants, one of the youngest women in the room, quickly replied: 'The big men in the company, they will only want educated women. If you are smart, educated – you will go to work for the mine. If not – you will stay in the village. Those working for the company will be alright. But me, I have no education. What will I do? This really worries me.' As a community, people in Paupe (men and women alike) were very self-conscious when it came to the generally low levels of formal education in the village. They worried that the company would prefer to employ other, better educated and more skilled people from further afield, leaving only a small pool of low-skilled jobs available to them. At the same time, they hoped that once the mine starts operating the company will open up schools in the Frieda area and create education and training opportunities that were lacking in Paupe – providing a solution to a problem the mine itself was expected to aggravate and contributing to broader development in the area.

Although many of my field conversations about the positive and negative impact of the mine spoke of the future, they were interwoven with remarks about the effect that the mine had already shown. At the most basic level, it has brought companies and state interests into the area and placed the Paiyamo in a series of expanding and opportunistic networks and introduced them to new kinds of powerful knowledge. It has exposed meaning and transformative capacities in old stories and unveiled a whole array of possibilities for the future. But it has also confronted some of the long-standing local ideas about land (see chapter 4), relations and personhoods.

Dangers of Jealousy

In the time I spent in Paupe, I observed, over and over again, ways in which sharing, care and reciprocity obligations played a crucial part in the local processes of community-making and provided underlying structures of the Paiyamo sociality. For example, kitchen doors had to be kept open when meals were being consumed as to avoid accusations of not wanting to share food. While I often heard women complaining about having to share their (often limited) supplies with anyone who came into the kitchen when a meal was being served, the same women would get offended if they heard that someone in the village had killed a pig and had not 'thought about them' when they were portioning and distributing the meat – always a long and complicated process in Paupe. Although the importance of sharing food and leaving kitchen doors open prevailed in the community, I recorded a few instances of the doors being closed at mealtime. On those occasions, store-bought food, not garden produce, was prepared and consumed. When I asked about the door being closed, the women circumnavigated the contentious subject, telling me: 'It's ok, tomorrow we will be eating food from our garden and the doors will be open.'

With the influx of cash into the community, communal dynamics – such as those related to food-sharing obligations – were being challenged. Money-earning opportunities played to the individual in people, while community requests and expectations constantly reminded them of the importance of the relationality and community obligations (see LiPuma 1998). This often left them internally conflicted or in open feud with other members of the community – frequently both. To give an example, a few members of the community had found employment with the Project and earned fortnightly wages; they had access to things such as the internet, phones and electricity, and they enjoyed Australian-style food and the relative comforts of the camp's residential facilities. Returning to the village at the end of their roster, they were expected to share their wages with members of their households and wider lineages. Claims and demands tended to also reach them at work, with other community members requesting their assistance with gaining access to company doctors, communication facilities and things such as medication, stationery, soap and newspapers. People who refused those claims, and those who chose to keep their wages in the bank and/or who denied requests for money, were openly accused of greed, criticised for neglecting their obligations and, on some occasions, publicly confronted for relinquishing their responsibilities to their families and to the community. Consequently, some of the workers would opt not to return to the village during their two-week break. Instead, they used the company's fixed-wing charter plane to travel to other parts of Papua New Guinea, where they would spend their money visiting family and friends, shopping, and pursuing a range of recreational activities. They would sometimes comment that doing so allowed them to spend their own money on what they saw as their own needs and wants and to get away (however temporally) from what felt to them like never-ending requests and expectations. They did not feel this necessarily made them bad or selfish as people and remarked, when asked, that they understood their obligations and responsibilities in the village but should be able to pursue personal interests and enjoy the benefits of their labour – despite negative comments such behaviour attracted from some members of the community. Similarly, people who successfully used their connections in Wabia and Ok Isai to gain access to the alluvial gold mining activities were expected to share their profits with their families and acquaintances in Paupe. They were not always prepared to do so. It was common knowledge in the village that, upon their return, such people would often try to use false stories of gold being stolen or dredging machines not working to avoid sharing their gold and/or money. Sometimes such tactics were effective; other times they were not, but all such stories raised suspicion, gave birth to rumours and often led to hostilities in Paupe. If caught out, the instigators who tried to deceive the community as to avoid their sharing obligations were shamed and expected to honour any pre-raised or subsequent requests. With the number of artisan miners in Wabia and Ok Isai growing, access to alluvial gold mining was gradually more restricted, with members of the two Telefol communities becoming increasingly hesitant to facilitate access for family members and the broader community. This led to significant tensions in the village, and between communities, especially when the need for money was greatest, such as at Christmas or the start of a school year. On one particularly memorable occasion, I observed a case in Paupe when a man forbade his wife from going to Ok Binai to look for gold, but she went anyway. As she was walking back to Ok Isai along a bush track, she was attacked by a wild pig. 'It was a large, black pig with big ears and long hair,' the woman said when she got back to Paupe later that evening. The attack left a large, deep cut in her leg. 'If the cut heals that means it was just an accident. But if it doesn't, that will mean that the pig was a bad spirit and that the attack was a result of *sanguma* [sorcery],' Dinah and Rhonda explained to me later and added: 'She should have obeyed her husband.' The cut did not heal easily. And so the following Sunday, at the end of the weekly service, the woman and her spouse came to the front of the church. Members of the community formed a circle around them – their hands stretched towards the couple – and prayed aloud asking for God's love and calling for His help against bad spirits, greed and jealousy.

According to local narratives, *sanguma* – a disruptive and destructive form of witchcraft where one person consciously uses dark magic against another – arrived at Frieda from the Sepik. During my fieldwork, I came across consistent claims that in the past, in the time of ancestors and *kastom*, the Paiyamo did not practise *sanguma* and that it only ever appeared in the community with jealousy resulting from changes in the local political economy. 'Before gold and the company there was no *sanguma* here. But now some people have money and others are jealous. So they adopted the ways of the Sepik and brought *sanguma* into Paupe,' I heard a number of times. 'Now this place is filled with jealousy,' Dinah warned me early on in my fieldwork and said that if people are jealous of someone, they can harm them. At the time, she was particularly worried about one man in the village who was a school treasurer and a youth leader, as well as a local tradesman and shopkeeper. 'People are jealous of him. They are watching him closely and talk about him behind his back. He needs to be really careful. This is why I asked Marvlen not to open a business in Paupe. It is best not to,' she said, referring to her husband and Mawi's dad. Many negative happenings in the village were attributed to jealousy and *sanguma*, like when rats destroyed the supply of flour in one of the shops in Aston, or when the engine broke down on a canoe of a man who helped the Sepik traders bring their petrol cargo up the river. On a moral level, *sanguma* was not an easy or a clear-cut subject for my hosts. They vocally condemned all instances of this kind of witchcraft, but more often than not they allocated the blame and responsibility to both – the person who used the magic to cause harm and the person whose actions somehow provoked it. I could not get rid of an impression that, in Paupe, *sanguma* served as a reminder of people's networks and obligations, keeping in check actions driven by predominantly individual rather than shared interests, particularly where this was done to the perceived detriment of the latter.

Interpreted within anthropological debates about Melanesian personhoods, *sanguma* emerged as a point of tension between individual interests and dividual aspects of local personhoods. It did not always target the person in the wrong but could affect someone that the person loved and/or nurtured. Increased instances of *sanguma* in Paupe were linked to growing opportunities and incentives for individual growth and financial wealth and the choices people made to hide their newly formed networks from others in Paupe, as to direct them for personal rather than communal gain. Tensions along similar cosmological lines were also starting to appear in community leadership. Although the variety of leadership styles and interests at Frieda cannot be underestimated, in very general terms, a distinction can be made along generational lines. Older leaders unswervingly spoke of the importance of *kastom* and used knowledge of *haus tambaran* to assert their position and, in the context of the Project, provide evidence of Paiyamo claims to land in the Project area. They responded to the instances of *sanguma* by reminding people of their shared ancestry and communal obligations and, engaging with the Project, prioritised recognition of rights to ancestral lands and shared interests in access to services and civic infrastructure. In conversations with the company representatives, the younger generation of leadership figures focused on pursuing contracted business opportunities and establishing a political and commercial representation body known in Papua New Guinea's resource contexts as Landowner Associations. While at the time of my research local youths had not yet reached the age that would grant them recognition and status associated with leadership positions, they often spoke about education and employment opportunities, through which they hoped to increase their professional capacities and pursue material wealth. Although the role and position of traditional leaders in the community were not being disputed, many of them, such as Tikap and Jacob, increasingly felt they had to defend their position and their knowledge, and they made an active effort to teach younger community members about the importance of community and *kastom* and about what it means to be a good person, a good Paiyamo.

Ideas of what it means to be a good person applied not only to community members. The Paiyamo often talked about the company and its on-site personnel in terms akin to those used to assess the behaviour of people in the village. For the Paiyamo, the promise of a prosperous future was inherently tied to the company's ability to deliver development, which in turn depended on the outcome and effect of complex and contested processes of stakeholder engagement. Their high expectations regarding their relations with the company were heavily informed by the legacy of past companies conducting exploration work at Frieda.

> The previous companies, they were good. Highlands Gold gave us a clinic and provided food and accommodation for all patients and their families. When someone died, they gave us coffee, food. If we asked them for a new motor, they would bring it on the charter. CEC gave all its workers store foods, matches, cigarettes, salt and soap, together with their fortnightly wages. The one after, too – if we wanted to travel on a company plane – there were never any problems. This was good.

Comments such as this one were a common feature of conversations about the Project's history and subsequent companies working at Frieda. 'It goes like this,' Thomas, a member of the Suna clan, explained to me one afternoon, 'Ground is important. So when *papagraun* ask the company for something they have to give it to us. In return, we are happy for them to use our land. We help them with access to the land, and they too must help us and give us what we ask for.' In many conversations, my interlocutors expressed their disappointment and sometimes even anger with Xstrata. 'This company now, they break the rule that was set up here a long time ago. That if we ask them for something they need to give it to us. But Xstrata, they only help us with serious problems. Everything else, they don't. But they should,' I heard at one of the community meetings in Aston. People's expectations and the ways they depended on the company for access to goods, services and – ultimately – development were further reinforced by the absence of government services and assistance. 'These are no longer the times when Australians were looking after us and when there was enough of everything for everyone. We should be happy we have the company here because the government has completely forgotten about us.' This was a common sentiment among my hosts. Such a narrative of dependency on the company for broadly conceived development assistance was actively encouraged by the government representatives during the encounters I observed in Paupe, laying bare the government's own expectations of the company's ability to deliver development to what it saw as some of the most geographically remote and hard to access parts of the country. As a result, the company and its on-site personnel were faced with the difficult task of managing expectations and controlling their liabilities without alienating the state and the Project's local stakeholders.

The Community Affairs Team

From the perspective of the Paiyamo, the company was enacted in the Project location by the Community Affairs team – company employees with whom they formed and maintained interpersonal relations. By definition and design, the team occupied a difficult and contested space between categorical prescriptions of the state and company management and operational systems and the reality of the Project's complex and multilayered daily operations, and they were given the trying task of enacting and managing both. The team was established when Xstrata took over the management of the Project in 2007. As described in the company's sustainability report for that year (Xstrata Copper 2008: 14), the team's role was to 'strengthen [the company's] relationships with landowners and other stakeholders of the Project and to better respond to their issues and concerns'. The team was tasked with fine tuning and delivery of the company's Community Development Plan; enacting the company's corporate social responsibility ethos and Community Relations policies; and enforcing the company's community relations policies in a way that was 'characterised by open and honest engagement with all stakeholders through effective, transparent consultation and cooperation' (ibid.). When it was first formed, the team consisted of a manager and three officers, who faced the task of 'building community relations from scratch', as none of the community relations personnel working at Frieda under previous management decided to stay on. Although on a personal level those relations were indeed new, the long history of the Project meant that in categorical terms the newly established team entered a highly complex and contested system of stakeholder relations that, by that point, had already acquired a difficult legacy.

The team grew over time and in 2011–12 consisted of three male superintendents, two of whom were expatriates, and seven to eight officers (a mix of men and women). Almost all of them were recent university graduates recruited and brought to Frieda from different parts of Papua New Guinea. None of the officers came from the impact communities, and only a couple of senior team members had previous first-hand experience of working at one of the country's already operating resource extraction projects. The team was led by a Brisbane-based Community Affairs manager, who made regular visits to the site. Upon commencement of employment, each team member received a long and detailed job description that was based on an equivalent role at an already operating mine owned by the company. Although the duties described in the document seemed familiar, team members (known locally as Community Affairs [CA] officers or simply as CAs) did not think the job description appropriately captured the nature of their roles at an advanced exploration project. There was also no job-specific induction for new team members and no set template procedure; as one of them put it, this left officers to 'figure things out for themselves'.

Team offices used to be in an old CA building at the Airstrip Camp. In 2011, the team was moved to a newly erected and well-equipped facility located in the Northern part of Piamoku, just inside the gate that separated the camp from its external environment. Members of the impact communities could contact and meet the CA officers with relative ease, and the team became an important part of the local social landscape. When not in the field and away from the camp, they tended to spend their working days inside the office. However, in the evenings – and much to the joy and amusement of the Paiyamo – some of its members could be seen catching fish on the banks of the Frieda River at the edge of the Airstrip Camp. In 2012, the CA team, together with the Project's medical personnel, was moved to the company's Base Camp where they could be reached by helicopter shuttle from the airstrip or, alternatively, a strenuous two-day walk through some very inhospitable terrain. This limited the ad-hoc interactions between the officers and local communities. The team members were not consulted about the move. One of them told me later that 'it was an operations decision. They told us we should be happy that we get less disturbed by the locals in the Base Camp. That we should enjoy. But that was opposite to our function'.

All of the CA officers I spoke to at Frieda linked their aptitude for their job to their ability to 'really get to know the communities', which they said they could only achieve by maintaining regular contact and building meaningful interpersonal relations with people living in the Project area. Through forming and maintaining those relations, they told me, they were able to get to know the people of Frieda in a way that was not possible in the company's meeting rooms and offices. Some of those relations were deeply personal, and many were based on actions and encounters that extended beyond the CA officers' roles as prescribed by their standardised job descriptions. The officers entered into their own care and reciprocity relations with community members, for whom they felt responsible, and they were genuinely affected by events and developments therein. For example, it was common for CA officers to spend a portion of their income on basic hygiene supplies or baby formula for communities. As one of them put it: 'We are the section of the company that really cares for the people and that takes on their concerns.' Frequent interactions were key to building and maintaining interpersonal relations with community members. CB radios were installed in all of the Project's impact communities when Xstrata took over management of the Project, and every day, at 8:00 AM, in the words of one of the CA officers, 'messages and stories were exchanged and relationships forged'. Team members visited one village a week on a rotational basis, and although at the time those were rather short, day visits, plans were being made to arrange for the officers to stay with communities overnight. 'When you sleep [in the village], it is different. In the evening, you can sit by the fire until late, and this is when you can hear the real stories, find out the real problems,' one of the team members told me. They would also frequently go to communities outside of scheduled visiting times – to tend to specific matters, such as village census; provide important project updates; or when dealing with medical or law and order issues that required prompt attention.

The CA officers spoke openly about what they perceived as the conflicting nature of their jobs, where their professional roles and capacities were at a constant interplay with the personal ties they formed with people in the Project's impact villages. Formally, they were meant to represent the company in the Project location. They had a contracted duty to Xstrata, and their work portfolio was geared towards advancing the company's interests at the Frieda River Project site. For many community members, the CA officers *were* the company, as the company often placed team members at the receiving end of communities' complaints and requests. This meant that the officers were held personally responsible for the company's decisions and would find themselves in a difficult position of dual accountability, caught between key performance indicators and internal targets and local regimes that made them personally accountable to community members in the Project location. When asked directly about their role in the Project, the CA officers consistently described themselves as 'bridging elements' and as 'mouthpieces' working between the Project's numerous stakeholders – positioning the team in a complex and challenging space between relations, roles and expectations. 'Our main role is to provide a link between the government, the communities and the Project. We go in-between. We need to make sure that the communities are aware of what the Project is doing and that we communicate any issues which arise in the communities back to the company.' This was a typical answer to my questions about the CA officers' role at Frieda.

In categorical terms, the role of the Community Affairs personnel was embedded in the company's ethos of sustainable development[4] and corporate social responsibility.[5] As described in its Sustainable Development Policy (Xstrata Copper, 2013), the company's goal of sustainable development was framed in terms of balancing social, environmental and economic considerations in how it managed its business, and its projects.[6] In achieving this goal, the company followed Xstrata Copper's 17 Sustainable Development Standards. These were aligned with several international standards including the UN Global Compact and the International Council on Mining and Metals sustainability principles and guidelines. The company then reported on its sustainability performance using GRI G3 Sustainability Reporting Guidelines.

Sustainability and Liability

Fixed temporalities of resource extraction projects and their inherently limited lifespan[7] played a role in the way the content of the reports spoke to the issue of sustainability in general and the company's aim to contribute to development of sustainable communities at Frieda in particular, promoting the company's aim 'to invest in skills, education, health, and social and enterprise development that lead to economic benefits and sustainable communities that outlive our operations' (Xstrata Copper 2010: 10). The company's sustainability reports for the Project (Xstrata Copper 2008–2013) talked about establishing community development programmes that would bring long-term, sustainable benefits to impact communities, conceptually merging community development with sustainable development. This focus on community development activities resonated well with the Paiyamo, who had very limited access to government services and looked up to Xstrata to bring development to the area. The company had a Community Development Plan. The plan was developed off site and in consultation with external experts. Once a year, the Community Affairs team at Frieda would hold a workshop during which they adapted the plan to the dynamic realities of the Project – allocating tasks, setting and checking schedules, and considering issues that they thought, on the basis of their knowledge of local context and relations with members of the impact communities, could affect the plan delivery. Team members told me that 'all the intelligence and little gossip gathered from visiting communities' were invaluable in this process. The plan gave them direction and set constraints on the team's activities. In one way or another, those constraints stemmed from two core factors, both of which carried a strong temporal element and were enabled by the way the company understood the mine at Frieda as a future and uncertain venture.

Firstly, the Frieda River Project was not an operating mine. This meant that it entailed a significant cost, not offset by any income. As the final decision regarding the future of the Project under Xstrata's management had not yet been made, the company placed strict limits on spending on community development activities. The CA officers were aware of the budgetary constraints and understood the need for financial caution, especially in light of high and not always realistic community expectations. They felt strongly, however, that the company must be willing to invest if it wants to build and maintain strong relationships with the Project's impact communities. They talked to me about the ease with which decisions to limit development assistance could be made from a distance of Brisbane offices and the difficulties of communicating and explaining those decisions locally whilst trying to maintain positive relations with communities and secure their ongoing support for the Project. In Paupe, refusing a request was seen as very poor form and a way to cut off rather than build and maintain relationships. Voicing his frustration, one of the team members told me: 'The company talks about maintaining the Social Licence to Operate.[8] But we also have a Personal Licence to Operate [that] we must maintain here.'

The second set of constraints was linked to what the CA officers described as the company management's use of the language of sustainability and social responsibility to justify their decisions to restrict community development programmes and involvement in delivery of infrastructure and services in the impact communities. In this interpretation, and with particular consideration for the uncertainties of the Project resulting from its 'undeveloped ore' and 'stranded asset' status, 'sustainability' meant that the company should not roll out any programmes that could not be sustained locally in the event of Xstrata's withdrawal from the area. The approach deemed such projects not only unsustainable but irresponsible. To quote one of the CA officers: 'If we introduce a service – will we continue providing it? And if we leave – can it be sustained? Everything we do, we must consider those questions. That is what they tell us.' For example, the CA officers told me that the company used the notion of sustainability to place restrictions on its support for the establishment of local schools. Company management considered education as the responsibility of the government and argued that a school that depended on support from the company was not sustainable in the event of their withdrawal from the Project or, for that matter, following the mine closure. After a conversation with a member of the management team, one of the CA officers explained to me: 'What we do needs to be sustainable. For example, we could pay the teachers – take this responsibility away from the government. The solution is immediate, at face value – but what about the long term?'[9]

Communicated from the company headquarters, this interpretation of 'sustainability' informed the way that many of the CA officers conceptualised sustainable development at Frieda. They would rarely speak of corporate social responsibility. When asked directly, they were most likely to respond by saying that, to quote two of them, it was 'something for the big bosses in their offices' and that it happened 'at a higher, management level'.[10] However, in our conversations, the CA officers would often refer to what they knew as irresponsible mining projects operating in other parts of the country. They talked about the issues with immigration at Porgera, the environmental degradation at Ok Tedi, or the breakdown of relations at Panguna on the island of Bougainville, and they were determined not to allow these stories to repeat themselves at Frieda. 'In some other projects, companies used the ignorance of the people to destroy the people. Here we are mindful of that and we do not want to do that. That would be disastrous for them. But it would be very bad for the company, too,' I noted one of them say. Whilst they did not talk about corporate social responsibility, they did talk about sustainable development, which they mostly saw in terms of community development. A large proportion of CA officers took sustainable development to mean the local capacity to sustain the attributes of development brought about by the Project following either the mine closure or the company's departure from the area should the Project not progress to the construction phase of its life cycle under Xstrata's management. As one of the team members told me 'Communities need to have the capacity to survive after the mine. Overdependence on the company is an ever-present danger. It's not sustainable, spoon-feeding landowners over and over. One day all this will come to an end and then, are you ready to stand on your feet?' This way of thinking made some, but not all, team members more understanding of the management's decisions to decline requests for development assistance in the impact communities. To quote one of them: 'It's challenging, you know, it makes it difficult at times at exploration stage. We want to do something that can be here tomorrow – this challenges us in [the] ways in which we do things today.'

Acting on Knowledge and Relations

From the perspective of the CA officers at the Frieda River Project, and on the basis of their considerable collective experience, two factors were critical in delivering effective and sustainable community development programmes, and thus securing access to land and the ongoing support of local stakeholders. Namely, the knowledge of and respect for local context, and the quality of interpersonal relations between members of the impact communities and the company's personnel. While the first entailed encounters with knowledge that did not easily translate into the technical language favoured by the company, the second required sustained engagement with and careful management of complex and constantly evolving social networks. To start with the latter, forming meaningful relations in a highly contested epistemological space became one of the most difficult and important tasks of the Community Affairs team. When it was first formed, the team's members were faced with a difficult legacy left behind by previous companies present in the area. This was characterised by high demands and expectations and a strong dependency on the company and its resources. Although I was unable to find any official documentation relating to the previous companies' community relations strategies, the stories the Paiyamo told of those early encounters painted a picture of a very handout-heavy company culture, where companies routinely honoured communities' requests for goods, money or assistance in return for undisturbed access to their lands. This form of engagement fitted closely with the business ethos of the mid- to late twentieth century, which largely excluded indigenous communities from any meaningful participation in the process of stakeholder dialogue. It was (and it could be argued that it still is) particularly symptomatic of resource exploration companies focused on securing short-term relations with local communities, often without much concern about how those will be managed in the long term. By the time Xstrata entered the Frieda River Project, and in the context of an increasingly popular and powerful discourse of sustainable development and corporate social responsibility, such short-term solutions were increasingly considered irresponsible, unfair and, ultimately, difficult to sustain. They were replaced with a long-term goal of building and maintaining meaningful stakeholder relations through interpersonal interactions aided by community relations and development programmes. Members of the Community Affairs team told me that they found personal relations the only means to manage communities' high expectations of development assistance.

Getting to know community members through direct, personal interactions had another effect on the CA officers. The more insights the officers were able to gain into the intricacies of social life in the Project's impact communities, the more uncomfortable they were with the generalisations used by the company and the government in an effort to get to know and build relations with the Project's local stakeholders. Especially as, in their view, the company had a tendency to understand local communities as a risk to be managed[11] and imagined them to be of a particular kind, lacking capacity, possessing little in terms of formal education and training, lagging behind progress and development and in need of the forward-looking mining industry. The officers were very vocal about the need to design engagement strategies and deliver development programmes tailored to local contexts, however challenging this might be, and defended this view on moral as well as business grounds. In the words of one of them: 'We want to make money. Our aim is to make profit. But if the development you are putting in place is not socially desirable, the investment will be destroyed, resulting in negative community reactions which will have negative consequences for the Project.' By spending time in local communities and forming relations with their members, CA officers developed competencies in various aspects of local *kastom*, albeit to varying degrees, including ideas about land and group affiliation. They said to me that they have found it difficult at times to act on this knowledge and their relations with the communities within the relative rigidity of the company bureaucracy, administration and hierarchical management structures. As I was told once during an interview with a CA officer: 'Community talk is often in line with their own beliefs. But the company has its own beliefs, and its own policies. It is an international corporation and it seeks to standardise policies and procedures. But this causes problems for us on the ground.'[12]

This particular CA officer clarified later that these concerns referred not so much to the company's own policies at the Project but externally set standards and indicators (including those enforced vertically by Xstrata Copper) passed down from the company offices in Brisbane and against which the team's work and the company's claims of good corporate citizenship were measured.[13] 'Our work should not be driven by ticking externally set boxes but by the work we do on the ground; specific community development goals that are set for this Project and for these communities,' I heard the officers say. When explaining the work of the team to me, one of them remarked that 'Community Affairs work is different to a pure study. It is not just about ticking boxes for feasibility. It is about maturing relationships for the future stages of the Project'. This is why, despite the bureaucratic demands of their jobs and the long hours required in the office in front of a computer, members of the team tried to spend a significant portion of their time with the communities and used personal relations to get to know the local people and what they themselves referred to as the local *kastom*. 'We try to find ways to understand the locals, their lifestyle, their thinking. We do this to find the right approach – one that will help the company create better relationships with the communities,' they told me. Although many admitted that they were often left puzzled by the intricacies of local knowledge practices and *kastom*, the senior team members, especially, voiced concerns over the risks posed by creating programmes based on cultural oversimplifications and misunderstandings, which, they told me, could threaten the long-term sustainability of stakeholder relations at Frieda. 'We believe that if you agree to embed structures in fundaments different than the traditional organising principles, this creates a serious threat for the future,' one of the CA superintendents explained. 'To do things well here we need to look inside the culture, and really engage with how the local people think,' added another. This last statement is symptomatic of an increasingly prominent industry-wide recognition that 'development initiatives that pay attention to local perceptions and ways are more likely to be relevant to people's needs and to generate sustainable interventions' (Sillitoe 1998: 224; see Gilberthorpe 2013b) as local indigenous knowledge brings to the table relations that science is trying to put aside (Kirsch 2004). In practice, however, such initiatives can be very difficult to achieve.

The CA officers at the Frieda River Project had found it difficult to act on their knowledge of local context and to make it visible and effective within the company structures. Reflecting on those difficulties, they put this down to two factors. Firstly, even though almost all of the team members I spoke to at Frieda saw their role as essential to the success of the Frieda River Project, many of them did not feel that the importance and the challenging nature of their role were fully recognised by the company's other departments, and they felt largely removed from centres of decision-making.[14] Unlike community members, many of whom considered the officers to be the company, the team members saw the company as a bigger and largely unfamiliar structure in which their role tended to be marginalised. At Frieda, they regularly found themselves in competition with the Project's operations division over access to financial, time and logistical resources and felt that it was 'difficult to do things when it is the operations who have the last say'.[15] Secondly, resource extraction is a data-driven industry that relies on particular pieces of information for decision-making. In order to meet the complex and often conflicting demands of their jobs, the CA officers had to use their knowledge to produce information that would fit into administrative categories and tick clearly specified boxes. Operating within the company's knowledge system, the work of CA officers was informed by the company's quest to categorise and organise its social responsibilities to make them manageable and auditable. In the process, they often found themselves having to use the same language and reproduce the very same ill-fitting labels and generalisations about local communities that they so eagerly critiqued amid a growing realisation that, to borrow from Jenkins (2004: 24), these 'did not necessarily fit the lived experience of the people in question' and were therefore unlikely to show the anticipated and desired effect. They were sceptical of the content of sustainability reports produced by the company, which used information requested by the management and provided by various team members, and they often saw it as preoccupied with presenting information that they felt did not fully reflect the reality of their experiences at Frieda. For example, while the information provided in the reports spoke of the *quantity* of engagement (by, for example, including a number of community visits and time spent on community development projects) alongside aspirational aims such as 'supporting community education' and 'developing capacity in communities' (Xstrata Copper 2012), the CA officers felt that the reports did not capture the critical knowledge about the actual *quality* of engagement and relationships between the company and its local stakeholders. Most of them were aware of the limitations this placed on the process of meaningful stakeholder dialogue but felt they had no power to change the situation. As one of them told me: 'Sometimes we know what the communities want and need. But to get it for them we need to ask the company, fill in forms, then wait for their approval. By the time we've finished doing all that, this request is no longer what it was. But this is the only way we can make development happen.'

The CA officers at the Frieda River Project felt conflicted between categorical prescriptions of their roles set out in employment contracts, and enacted within a broader and largely unknown company structure, and interpersonal relations they formed with people in the Project location. To an extent at least, this tension stemmed from what they saw as the company's ideas about corporate social responsibility framed around the notion of managing risks to the Project, which contrasted with the sense of personal responsibility the officers felt towards members of the impact communities at Frieda. The company's understanding of the mine and its temporalities played a large part in the process, as management's decisions were judged on-site in terms of limiting liabilities and carefully controlling how the company entangles itself in local contexts, as to be able to exit the local nexus of relations at relatively short notice should it decide to leave the Project and not take it through to construction and operation. To that end, the company employed a targeted interpretation of the ethos of sustainability to manage its liabilities whilst simultaneously claiming that these actions were driven by and demonstrative of its claim of good corporate personhood. The CA officers, on the other hand, performed their roles by entering into a dynamic nexus of interpersonal relations embedded in local temporalities and ideas about knowledge and relations, with all the rights and obligations this involved. They felt that in the event of their departure from Frieda those interpersonal connections would be much more difficult to dissolve, and anticipated that personal obligations would continue long after contractual ones have expired. In daily conversations with community members, the CA officers would not commonly discuss the uncertainties characterising the future of the Frieda River Project – although these no doubt played a part in shaping their categorical roles and impacted their activities. However, some of the members talked about uncertainties relating to their own positions at the Project. They came to see their own knowledge of local context and *kastom* as embedded in people, places and relations and worried about the consequences their departure might have for stakeholder relations at Frieda.

Making Development Happen

Early on in this book I pointed to the limitations of stakeholder models and argued that, put into practice, all categorical prescriptions are filled with tensions and contradictions. I described this phenomena as dissonance within stakeholder categories and argued that it was to be found in the dynamic spaces between categorical prescriptions of stakeholder roles and responsibilities and their enactments in the Project location. I also described a second type of categorical dissonance, one that emerges between categories, where one or more stakeholder groups fail to meet the expectations of their role, leading to absences in their respective functions. I argued that where such absence threatens the shared project another actor (or actors) with the capacity to do so will likely fulfil the missing role to address the disharmony and ensure continuity of the process – in this case, moving the Frieda River Project towards construction and operations. In one way or another, both types of dissonance can be attributed to all key categorical sets of relations at Frieda, where the landowner label was used to describe a diverse, ever evolving and largely imagined social form and where the company was enacted by people who felt they had little knowledge of the company as a bigger entity and limited influence over its decisions and actions. What I want to focus on here is how a phenomenon I refer to as the absent presence of the state has revealed dissonance between stakeholder categories of the Frieda River Project, forcing the company to take on what, in prescriptive terms, were considered responsibilities of the state in order to resolve a disharmony that, it judged, had the potential to pose a serious risk to the Project.

I use the term absent presence of the state to refer to a phenomenon in which the presence of the state is felt through its absence and people's experience of the gaps this absence creates. In the context explored here, these gaps appeared between what could be termed as the categorical presence and interpersonal absence of the government from the local network of relations at the Frieda River Project. In a categorical sense, the presence of the state as a legislator was very much felt at the Project, its power demonstrated through legal prescriptions, forms, laws, procedures and its capacity to, when the time comes, grant the Special Mining Lease for the Project. Company personnel would often bring the state into conversation to explain its decisions and actions; for example, when justifying the amount of compensation payments due to communities in exchange for disruptions or damage to their land. But this presence was not equal to a relationship. In view of the Paiyamo, it lacked the interpersonal component and was devoid of the access to government representatives with whom consequential relations could be leveraged and built.[16] Those interpersonal absences – a lack of direct interaction with government officials – contributed to the feeling that the state was absent from Frieda, exacerbated by the near-absence of government-provided services and civic and transport infrastructure. Community members' very limited understanding of government processes and systems further added to this sense of alienation and absence, leading to high levels of frustration among the Paiyamo, who, faced with government regulations and bureaucratic prescriptions that were inflexible and non-negotiable, felt they could not really get to know the government.

Local ideas of personhood play a critical part in those processes. In cosmology that treats knowledge as an inherently relational endeavour, the Paiyamo interpreted the government's lack of interest in entering into interpersonal relationships with members of the community as evidence of the state's unwillingness or lack of interest in getting to know them. The Paiyamo judge knowledge and relations against the transformative effects these have on the world. Over time, my hosts in Paupe came to understand the anticipated mine at Frieda as an effect of knowledge and relations, which they actively navigated hoping to precipitate mining development in their area. This is why they were openly annoyed with the almost exclusively categorical presence of the state, which they did not know and felt did not want to know them, and with which no two-way relationship could be developed. In the absence of interpersonal relations with the government, members of the Paupe community did not identify with it in the way they came to identify with the company developing the Project, and they did not form relations with government representatives like they did with members of the company's Community Affairs team. This meant that they felt less able to affect change through interacting with the state, focusing instead on engaging with the company and its representatives in the Project area. This is not to deny a strong categorical relationship between the Paiyamo and the company, grounded in PNG legislation and prescriptions of legally recognised customary land ownership. In this relationship, the company (the developer) had to secure and maintain access to land owned by the Paiyamo (the landowners) in order to develop infrastructure essential for the Project's activities and future operations (see Bainton and Banks 2018). However, to my hosts in Paupe, it was the interpersonal relations they entered into with the company's on-site personnel that made the categorical relations meaningful and awarded them with effect – with the ultimate goal of leveraging those relations to see a mine at Frieda.

My hosts in Paupe were very vocal about their disappointing and frustrating engagements with the provincial and national governments (or rather lack thereof) – institutions they viewed as powerful but ambiguous and with whom they shared passing and limited relations at best. For all intents and purposes, the way they saw it, they did not have a relationship with the government independently of the Frieda River Project. While they were always keen and excited to talk to the government representatives during their rare visits to the area, they felt very little obligation towards the state and looked instead to the corporation for what would traditionally be seen as the roles and responsibilities of the government. 'We must support the company and work with them. It is the company that will bring services to us. It will not be the government who will come and give medications to our children. If the company was not there, the government would not remember about us,' I heard one of the leaders instruct the community during a village meeting. The rhetoric of 'people forgotten by the government' featured strongly in such statements, and the Paiyamo turned to the company for goods and services that they came to associate with a promise of development that they had read from the imagery of the anticipated operating (and profit-making) mine on their lands. In the few encounters I witnessed between the Paiyamo and government representatives, the latter actively encouraged the communities' dependency on the company for both civic infrastructure and basic services. During his brief visit to Paupe, I heard a member of the provincial administration say to the Paiyamo community: 'Before, you all lived in the bush and you had nothing. The government is too far away, it will not help you. If it was not for the company you would still be in the bush with no development, nothing. If you don't support the company, you will go back there.'[17]

Rather unsurprisingly, the company representatives at the Project felt uncomfortable with this build-up of expectations and the pressure they found themselves under to fulfil what they saw as the role of the state. 'The government doesn't play its role. It thinks that since we are here we can provide everything,' one of the CA officers told me following a community meeting in Paupe. 'But we cannot be seen as the provider of government services in the area – this is just not right,' he added. While the CA officers, as well as the company's higher tiers of management, intensely resisted taking on what they saw as the role of a quasi-government, they often felt that they did not have much choice if they wanted to maintain good relations with the impact communities and secure and sustain their consent for the Project. 'The government thinks that we can solve all its problems'; 'The government is always two steps behind, not keeping up the pace'; 'The government relies too heavily on the company to provide development assistance. It is not taking initiative'; 'The government never played its role properly. As a result, the communities, too, are not being proactive with alternatives. They are just waiting for the mine.' These are just some of the company voices I heard expressing frustration with the lack of active state involvement in providing services and development assistance to Frieda's communities. When I spoke with the provincial government representatives during one of their visits to the area, they themselves admitted openly that they depended on the company's resources and aptitude to bring development not only to the six impact communities but also to the two affected provinces. This anticipation was shared by the national government. As a Mineral Resources Authority representative explained to me: 'The national government has one hope – for the project to develop as a mine. We are asking, preaching, hoping – when will it happen? When will the company submit the SML application? We are waiting [and are] happy to grant the lease. We are waiting for the economic impact and development.'

Many of the conversations I had with the CA officers about the lack of government involvement in delivery of infrastructure and services at Frieda, and its dependency on the company to fulfil some of its own roles and responsibilities, spoke of a lack of capacity to deliver development at Frieda rather than its bad intentions or lack of willingness to do so. In some conversations, my interlocutors linked this to the weakness of the national government passed down through all levels of state governance. 'The governments know that it is their job, but they lack the resources, the manpower. They always thank us. They are very appreciative. They know that we are doing what they should be doing. They are always thankful,' one of the CA officers explained to me. 'But they lack the capacity and resources and as a result the company finds that it must provide services it really shouldn't have to. And then there is the issue of sustainability. If the government is not involved, no one will look after what we put in here. This will cause big problems in the future.' Other company employees were less sympathetic. They were of an opinion that the government purposefully took a step back and let the company fill the gap in providing services and development assistance to the local communities. 'The [company] feels compelled – it has no choice,' I heard from one of them. As I described above, the company management used the ethos of sustainability to push back on requests for development assistance where they felt these encroached on state competencies and the government's obligations to its citizens. On occasions, however, the company did step in; for example, by installing water tanks to allow communities access to clean drinking water; by providing transport for government-appointed teachers; and, in some cases, supporting teachers financially when the government failed to pay their wages. The company did not do this because it intrinsically wanted to, but because this allowed it to maintain the support of local stakeholders and advance its own main role at Frieda as that of a commercial business and Project developer. It saw its function with regard to community development and engagement programmes as an 'extension' rather than being 'in place' of government provision of basic services. In attempting to mediate the failures of the state and its absence in the Project location, the company addressed a disharmony that could potentially threaten the process of precipitating a mine at the Frieda River and expose the shortcomings of stakeholder models in contested fields of resource relations. As a result, while people enacting stakeholder categories in the Project locality continued to face ongoing tension and conflict over allocation and fulfilment of roles and responsibilities, the demands of the categorical model were met and the project continued to move forward towards the shared desired effect – the Frieda River Mine.

Conclusions

In his analysis of Melanesian temporalities, Gellner (1964) placed the concept of neo-episodic time in opposition to the so-called evolutionary time – where social change occurs through a gradual and continuous process. One of the key characteristics of the latter was that it presented time as inherently embedded in the ideology of progress and advancement, which placed excellence exclusively in the domain of the future. Gellner argued that, as a result, taken in the evolutionary view, time ceases to be morally neutral.

Looking at ways in which the government, the company and the impact communities conceived of a mine at Frieda as 'desirable progress', it may be argued that, tied to the idea of development, the mine itself ceases to be morally neutral. While the company used a range of technocratic mechanisms and indicators to make a financial argument for the Frieda River Mine, the Paiyamo sought to understand the mine using their own moralities, embedded in and brought to life through an intricately woven fabric of emplaced social relations, embodied knowledge and stories told of important events, what Phillip Schorch and Arno Pascht (2017) termed 'critical junctures' – moments in time where new relationships allow new paths to open up. In local cosmology, such junctures could be rediscovered in past events, as powerful stories of happenings and relations could be altered, and as new events and interactions revealed in them new meanings and novel opportunities for action.

Ethnographic material presented here illustrates some of the challenges associated with negotiating and creating shared frames of reference and people becoming known and knowers in contested processes of preparing for a large-scale resource extraction project. Experiences of the Community Affairs team highlight the positional peculiarity of those roles and illustrate the way in which the company understood a mine at Frieda as a future and uncertain prospect, which in turn shaped the way they understood and enacted their responsibilities towards the local stakeholders. This stands in contrast to the stories told by the Paiyamo about the effects the mine had already started to show, demonstrating divergent perspectives and interests, embedded in ways of experiencing and being in the world. In the time I spent with them, I observed the ways that the Paiyamo tried to manage the effects of a mine that, I was explicitly told by the company representatives, was not there; and how, by investing their hopes in the efficacy of stakeholder roles and relations, they enacted a belief in the transformative effect of mining development at Frieda. The long history of the Project, traced in local narratives all the way back to the time of ancestors, meant that they felt strongly invested in the process and were eager to see the mine they have long been waiting for. At the time this research took place, most of the people living in Paupe had been born after the first company arrived on the banks of the river, and they were brought up with the presence of the Project.

The chapter described a range of hopes and aspirations invested in the process of precipitating the mine at Frieda. There can be little doubt that the shared anticipation of development embedded in a belief of the transformative qualities of a large-scale resource extraction project played a crucial role in bringing the different stakeholders of the Project together and awarded traction to the process of stakeholder engagement, bringing with it new ideas about personal, collective and national viability (see also Bainton

2011). As mentioned in chapter 1, the term 'developing Frieda' spoke to divergent visions of what development entailed for the Paiyamo, the company and the government. But it offered a narrative of change able to unite people and groups representing very different cosmologies in a shared interest. In the process of stakeholder engagement, these people came together to shape their futures, working towards their imageries of an in(visible) mine at Frieda River.

Notes

1. In a similar vein, Golub (2006: 274) described how, as a direct result of their relations with the mining company and local concepts of development, the Ipili at Porgera 'conceive of a past as full of want, they strive for a future full of plenty'.
2. A few additional sources offer valuable insights into these processes in other parts of Papua New Guinea. Of particular interest might be Lohmann's (2000, 2001, 2003) description of the ways in which the Asabano relied on a Christian God to face what they perceived to be dangers of the 'outside world'; Robbins' (1998b:113) analysis of how for the Urapmin 'their Christian identity linked them to a community they imagine is far more successful both morally and materially as the national one' (see also Robbins 1995 and 1998a); and Jacka's (2002) insights into the processes through which the Enga and the Ipili fashioned their own version of participating in global events by linking stories described in the Bible to regional events. See also Hirsch's (2001b) study on the origins of modernity among the Fuyuge and, for a more theoretical contribution, Bialecki et al.'s (2008) review of how anthropology has linked the 'liberating' influence of Christianity to the conditions for capitalist economy and political modernity.
3. For an elaboration on knowing through hearing and knowing through seeing among the Telefomin, see Jorgensen (1981).
4. The most commonly quoted definition of the concept emerged from a report prepared by the United Nations' World Commission on Environment and Development and defined sustainable development as 'development that meets the needs of the present without compromising the ability of future generations to meet their needs' (WCED 1987: 43). Although over the years the phrase 'has been used to mean different things to different people in different contexts' (Bebbington 2001: 129), the increasing popularity and importance of the idea has opened up a crucial and desperately needed space for questioning what constitutes social progress and has led many to review the character of the dominant model of development (see Lélé 1991; Baker 2006).
5. In the absence of a single definition, corporate social responsibility has been described loosely as 'a bundle of practices and actions that take into account the expectations of diverse stakeholder groups and the triple bottom line of economic, social, and environmental performance' (Hunoldt et al. 2018: 2).

Little about corporate social responsibility remains uncontested or incontestable (Gray et al. 1995). Similarly to sustainable development, it has been operationalised and interpreted differently by different people. Its actual impact on business prospects and on people's lives remains largely unknown (Votaw 1973; Blowfield 2007), and the rhetoric itself has been widely critiqued as a powerful means of self-regulation that both avoids external controls and simply measures its own measurements (e.g. Gray et al. 1996; Adams 2004). See Carroll (1999) and Jenkins (2005) for the history of the concept. For studies of CSR from an anthropological perspective, including several case studies from the resource extraction sector, see Dolan and Rajak (2016).

6. The document (Xstrata Copper 2013: 20) stated the company's position that: 'operating to leading standards of health, safety and environmental management, contributing to the development of sustainable communities, and engaging with stakeholders in two-way, open dialogue . . . enhances corporate reputation and is a source of competitive advantage' and declared the company's intent to use these mechanisms to 'gain access to new resources, maintain a licence to operate, attract and retain the best people . . . and optimise [its] management of risk'.
7. The resource extraction industry, conditioned on the depletion of non-renewable resources, is often considered in contradiction to the ideology of sustainability (see Hilson 2001; and Tilton 1996).
8. Although it adapts technical and tangible language associated with licensing and permitting, in reality social license to operate is 'an intangible construct associated with acceptance, approval, consent, demands, expectations and reputation' (Parsons et al. 2014: 84). It has become a popular shorthand used to describe the way that companies deliver benefits to communities in exchange for the latter not interrupting their local operations. In academic debates, the term has been criticised on the grounds of limiting, not expanding, critical debates about entanglements and regimes of responsibility and accountability between companies and society (Owen and Kemp 2013; see also Owen 2016).
9. In recognition of prevalent power inequalities between stakeholder groups and their respective ability to gain leverage in decisions and processes that affect them (Bendell 2005; Adanhounme 2011; see also Dahl 1961), academic and practitioner voices increasingly advocate that the state must be actively brought back into the corporate social responsibility equation and that it must grow its capacities to effectively embrace its roles and responsibilities (Frynas 2005; Genasci and Pray 2008: 58; see also Kemp 2010; and Banks 2008) in order to help deliver responsible and sustainable development.
10. The company's sustainability disclosure used the notion of risk rather than intrinsic responsibility to provide a rationale for its engagement with the ethos of corporate social responsibility and was grounded in the company's belief that 'operating responsibly and to the highest level of international standards

mitigates risks, creates opportunities and enhances reputation and competitive position' (2008: 5).

11. This sentiment is common in literature on mining and corporate social responsibility. See, for example, Frederisken (2018) on corporate social responsibility as a risk management mechanism.
12. To give an example, in many parts of Papua New Guinea, the industry's emphasis on financial rewards and benefits as a means of promoting sustainable development has been widely criticised for trivialising relations and processes constituting the fabric of social life for the communities affected by mining projects (Burton 1997; Jackson 1997; Power 1997; Strathern 2004; Banks 2012 [1999]; Gilberthorpe 2013b). It has also been criticised for focusing on its own targets while disregarding local goals and local versions of development (Connell and Howitt 1991; Gilberthorpe and Banks 2012).
13. See Kirsch's critique of the concept of corporate personhood – particularly his criticism of a mining company, which, in his words, 'reveals a fundamental contradiction in the metaphor of corporate personhood: corporations selectively claim the rights of natural persons while ignoring the social responsibilities of personhood' (2014b: 211).
14. An industry-wide trend as demonstrated by Owen and Kemp (2014). While community relations and development personnel play an increasingly important role in resource extraction projects, they tend to have limited access to resources and decision-making power.
15. This is not unique to the Frieda River Project. While the discourse of corporate social responsibility has forced companies to hesitantly acknowledge that in an era driven by the rhetoric of sustainability good relations with their local hosts have become 'core to their business', they are not seen as their 'core business' and so operations divisions continue to maintain an upper hand over community affairs for most mining projects (Kemp and Owen 2013; 2015).
16. See also Philip Abrams' (1988 [1977]) work, which distinguishes between 'state-system' and the 'state-idea'.
17. The situation described here is by no means unique to the Frieda River Project. All around Papua New Guinea, resource extraction ventures at all stages of development operate 'against a backdrop of bad governance and ever-growing community expectations' (Director of the Papua New Guinea Chamber of Mines and Petroleum, personal communication; see also Power 1997; Toft 1997; and Jackson 1991), implicating corporate and state agencies in a game 'in which each occasionally tries to hide behind the other, but the steady diminution of the government's authority and capacity means that its own shadow keeps getting smaller and the companies feel more like shadow states' (Filer et al. 2008: 171). As a result of these contested stakeholder relations, over the years mining companies in Papua New Guinea 'have become increasingly involved in "community development" roles, often promoting the newly acquired cor-

porate language of "sustainability" and "partnerships"' (Banks 2008: 24). These processes are symptomatic of a much wider debate about the role of companies in development. Changes in the global and regional balance of influence resulted in reduced confidence in the role of the state to deliver development and successfully promoted the corporate sector as a new dominant source of social improvement in many parts of the world, including Papua New Guinea (see, e.g., Newell and Frynas 2007; Reed and Reed 2009; Kemp 2010; Blowfield 2012).

Conclusion

Located in a geographically remote area of Papua New Guinea, and home to one of the biggest undeveloped copper and gold deposits in the Pacific, Frieda was a place where indigenous ideas of knowledge and personhood collided head on with company notions of social responsibility and divergent ideas about managing a large-scale resource extraction project. Frieda River was and remains a 'complex orebody', where this complexity includes a range of environmental, social and governance factors. My aim in this book was to enhance understanding of those complexities by telling a story of local stakeholder strategies on the eve of industrial development, told from the perspective of the Paiyamo.

When I arrived on the banks of the river, I entered a distinctive and diverse community of people brought together by a shared imagery of future opportunities, framed around powerful visions of a large copper and gold mine. I observed with great interest an intricate interplay of categorical and interpersonal relations at the Frieda River Project and was intrigued by the ways in which my hosts and interlocutors strived to bridge and populate a gap emerging between what they saw as requirements of stakeholder categories and their experience of stakeholder engagement. The Paiyamo were not passive observers of these processes. Captivated by the imagery that linked the mine to a promise of a better future, they transformed themselves into 'a kind' and developed a range of strategies in an attempt to effect and direct anticipated change. Despite decades of exploration, there was no mining infrastructure at Frieda and no industrial resource extraction activities were taking place. However, from the perspective of the Paiyamo, the mine has already begun to have an effect on people and places and open up new pathways for moving towards what they imagined as a more prosperous future. The book describes how the Paiyamo made conscious and strategic efforts to understand, manage and precipitate those effects and the mine at Frieda River, embedded in their ideas of knowledge, belief and personhood and a local aesthetic of revelation.

Understanding what 'a mine' is in these contexts, and what it means to different people, is critical to recognising and comprehending the complexities of the Frieda River Project. I strived to demonstrate in these pages how a change in perspective makes us re-evaluate how we think about people, the world and the different elements within it. By following the Paiyamo in uncovering the history and telling the story of the Project, I sought to challenge how we conceive of 'a mine' in the context of resource exploration ventures. Pivoting on Strathern's (1988) theorisations about Melanesian personhoods, I set out to explore what 'a mine' would look like if we took something other than the physical form of a mining pit and industrial infrastructure as a starting point of analysis. My aim was to create within these pages a discursive space through which the mine at Frieda could be brought to life and made visible. To achieve this, consecutive chapters in this book traced the ways in which the Paiyamo came to know and understand the mine, and themselves in relation to the mine, and how they experienced it in terms of social relations, community and individual aspirations and novel mechanisms for affecting an array of future opportunities.

My ethnographic work at Frieda led me to pose a seemingly simple question that proved surprisingly difficult to answer: in the absence of mining infrastructure and industrialised mineral extraction activities, what is 'a mine' at Frieda River? For the company representatives, Frieda River was not a mine but a resource extraction project at an exploration stage of development. Referring to the Mineral Resources Authority's mining project life cycle, they told me that the project becomes a mine when a full feasibility study is concluded, the mining licence granted, and when construction of mining infrastructure allows for resource extraction activities to begin. When I posed the same question to my hosts in Paupe, I got a great variety of different answers. The Paiyamo would tell me the myth of *bangri* and the story of the German woman. They would tell me about the time when the first company arrived on the banks of the river and about how Aiyapanai nodded when asked if they could use the Paiyamo land and places for mineral exploration. They would tell me about the *haus tambaran*, the *masalai* and about how they welcomed the Holy Spirit and became Christian. They talked to me about how they became the Paiyamo and transformed into landowners and how their relationships with the company and the stakeholder engagement process revealed new meanings and new capacities in old stories and past events. They described the ways the mine has already shown effect and how copper and gold have brought different kinds of people into their area, providing the Paiyamo with access to new, powerful relational networks, new kinds of knowledge and ways of making it visible, and new opportunities for the future. They also explained how money and jealousy brought *sanguma* into Paupe. When I told pastor Jacob what the company employees told me, that there was no mine at Frieda, he looked down, shook his head and said: 'Don't mind them. The mine is here, *em stap* [lit. it is in place, it exists]. It just has not yet revealed itself.'

Jorgensen (2014: 33) argued that 'mining stories are never just about mining'. It took me a while, but I eventually realised that the stories of the German woman, the arrival of the first company and many others I heard on the banks of the Frieda River were stories the Paiyamo were telling me about the anticipated operating mine at Frieda River all along. When telling their stories of the Project, the Paiyamo often seemed unusually calm and confident in their position as stakeholders of a mining project and trusting in the imagery of the lives that will be lived once the mine starts operating. They were able to do so because, in many ways, for the Paiyamo the mine at Frieda was already happening but it had not yet revealed itself. From the early encounters with the mining industry, they came to see the mine as a kind of 'volition that makes things happen, but is not always visible' (Lohmann 2003: 183), a perspective that was grounded in local ideas of knowledge and relations and constantly reassessed on the basis of the effect the industry's presence was having on their world, amidst acknowledgement that this may never be fully known.

The Paiyamo were increasingly aware of the negative elements of the change brought about by the industry's presence in the area (as demonstrated, for example, by instances of jealousy and *sanguma* within the community). However, these negative experiences did not distract them from the promise of a mine, development and future prosperity. They told me numerous times that they were becoming very tired of the seemingly never-ending exploration. They wished that the company would stop 'checking everything over and over again' and that the government would 'hurry up and issue a mining licence'. 'This is what we want – before we die, we must see this mine and we must see development here,' many of them told me at different points in my fieldwork. They spent a lot of time talking to me about the mine and their lives in relation to the mine, eager to see its full effect. As one local woman told me: 'I was a little child when the company came. Time went by, I got married, I had children. Now all my children are married, too, they are working, having children, and I have not yet seen this mine. I do not know, will this change, or not?'

I have been very conscious in this book not to create here a false and unjust impression of the Paiyamo and Melanesian societies more broadly as having been ahistoric and somehow 'pristine' up until what is frequently termed in colonial contexts as 'the first contact'. Equally, I was mindful of the need to convey the magnitude and scale of change brought about by new ideas, new sources of power and leverage and new types of knowledge introduced into local networks by broader economic, political and ideological factors (and actors) linked to the industry and the powerful ideas of progress and modernity. When during my fieldwork the company announced an extension to the feasibility study and a delay in filing a mining licence application, people gathered in the wind houses and kitchens of Paupe and spoke about the significant worry and anxiety this news had brought to the village. These feelings were reinforced further when a rumour started that Xstrata had put its shares in the Frieda River Project up for sale. The Paiyamo have seen this a number of times in the past – subsequent companies winding down, packing up and leaving the area, only to be replaced by new ones, which, they told me, tended to begin with existing information being thrown out and questions asked regarding the people, the deposits and the area from the beginning. This time they feared that the blow would be even greater, as Xstrata had managed to carry the Project through to a full feasibility study – an advanced stage of the exploration phase of a life cycle of a mineral project. Despite the fact that the company often angered the Paiyamo by refusing their requests, and frustrated them with its seeming inability to understand the meaning of local knowledge and *kastom*, many of them trusted the CA officers. They believed that Xstrata would look after the communities and that it would operate a good and prosperous mine in the area. 'Whatever little money we have here now is thanks to the company. Whatever little service – the same. If the company goes, our lives will be rubbish again. We will be back in the bush, living like our ancestors. But we have come such a long way since their time – we really do not want to go back,' the local councillor told me.

I can count on one hand the number of times I heard the Paiyamo talk about what their lives would be like after the mine closes down. Although they were fully aware of the finite life span of resource extraction projects, they tended not to speak about the closure of a mine and the end of its life cycle. The one time I talked about it to one of the local elders he smiled at me and said: 'When the mine is finished, and after we die, this is what our children will do. They will find their own minerals, too, their own deposits. Or – they will go to a big school in the city, they will become educated and then they will go to work at another resource extraction project.'

References

Abrams, P. 1988 [1977]. 'Notes on the Difficulty of Studying the State', *Journal of Historical Sociology* 1(1): 58–89.

Adams, C. 2004. 'The Ethical, Social and Environmental Reporting – Performance Portrayal Gap', *Accounting, Auditing and Accountability Journal* 17(5): 731–57.

Adanhounme, A. 2011. 'Corporate Social Responsibility in Postcolonial Africa: Another Civilizing Mission?', *Journal of Change Management* 11(1): 91–110.

Allen, M. 2013. 'Melanesia's Violent-Environments: Towards a Political Ecology of Conflict in the Western Pacific', *Geoforum* 44: 152–61.

Anderson, A. 2011. *Landscapes of Relations and Belonging: Body, Place and Politics in Wogeo, Papua New Guinea*. Oxford: Berghahn Books.

Anderson, B. 1991 [1983]. *Imagined Communities: Reflections on the Origin and Spread of Nationalism*. London: Verso Books.

Anderson, G. 2008. 'Frieda Will Not Be an Ecological Disaster' (letter), *The National*, 30 May.

Arendt, H. 1958. *The Human Condition*. Chicago: University of Chicago Press.

Bainton, N. 2008. 'Men of Kastom and the Customs of Men: Status, Legitimacy and Persistent Values in Lihir, Papua New Guinea', *The Australian Journal of Anthropology* 19(2): 194–212.

———. 2009. 'Keeping the Networks Out of View: Mining, Distinctions and Exclusion in Melanesia', *Oceania* 79(1): 18–33.

———. 2010. *The Lihir Destiny: Cultural Responses to Mining in Melanesia*. Asia-Pacific Environment Monograph 5. Canberra: ANU E Press.

———. 2011. 'Are You Viable? Personal Avarice, Collective Antagonism and Grassroots Development in Melanesia', in M. Patterson and M. Macintyre (eds), *Managing Modernity in the Western Pacific*. St Lucia, QLD: University of Queensland Press.

Bainton, N. and G. Banks. 2018. 'Land and Access: A Framework for Analysing Mining, Migration and Development in Melanesia', *Sustainable Development* 26(5): 450–60.

Bainton, N., Ballard, C. and K. Gillespie. 2012. 'The End of the Beginning? Mining, Sacred Geographies, Memory and Performance in Lihir', *The Australian Journal of Anthropology* 23(1): 22–49.

Bainton, N., C. Ballard, K. Gillespie and N. Hall. 2011. 'Stepping Stones Across the Lihir Islands: Developing Cultural Heritage Management in the Context of a Gold-Mining Operation', *International Journal of Cultural Property* 18: 81–110.

Baker, S. 2006. *Sustainable Development*. London: Routledge.

Ballard, C. 2014 'Oceanic Historicities', *The Contemporary Pacific* 26(1): 96–124.

Ballard, C. and G. Banks. 2003. 'Resource Wars: The Anthropology of Mining', *Annual Review of Anthropology* 32: 287–313.

Banks, G. 2002. 'Mining and the Environment in Melanesia: Contemporary Debates Reviewed', *The Contemporary Pacific* 14(1): 39–67.

———. 2008. 'Understanding "Resource" Conflicts in Papua New Guinea', *Asia Pacific Viewpoint* 49(1): 23–34.

———. 2012 [1999]. 'The Economic Impact of the Mine', in C. Filer (ed.), *Dilemmas of Development: The Social and Economic Impact of the Porgera Gold Mine 1989–1994*. Canberra: ANU E-Press.

Banks, G. and C. Ballard (eds). 1997. 'The Ok Tedi Settlement: Issues, Outcomes and Implications'. National Centre for Development Studies, Research School of Pacific and Asian Studies, Canberra: Australia National University.

Barker, J. 1992. 'Christianity in Western Melanesian Ethnography', in J. Carrier (ed.), *History and Tradition in Melanesian Anthropology*. Berkeley: University of California Press.

———. 2010. 'The Varieties of Melanesian Christian Experience: A Comment on Mosko's "Partible Penitents"', *The Journal of the Royal Anthropological Institute* 16(2): 247–49.

Barth, F. 1975. *Ritual and Knowledge among the Baktaman of New Guinea*. New Haven: Yale University Press.

———. 1987. *Cosmologies in the Making: A Generative Approach to Cultural Variation in Inner New Guinea*. Cambridge: Cambridge University Press.

———. 1990. 'The Guru and the Conjurer: Transactions in Knowledge and the Shaping of Culture in Southeast Asia and Melanesia', *Man*, New Series 25(4): 640–53.

———. 2002. 'An Anthropology of Knowledge', *Current Anthropology* 43(1): 1–18.

Bateson, G. 1958. *Naven*. London: Wildwood House.

Bebbington, J. 2001. 'Sustainable Development: A Review of the International Development, Business and Accounting Literature', *Accounting Forum* 25(2): 128–57.

Bell, J.A. 2009. 'Documenting Discontent: Struggles for Recognition in the Purari Delta of Papua New Guinea', *The Australian Journal of Anthropology* 20(1): 28–47.

———. 2014. 'The Veracity of Form: Transforming Knowledges and Their Forms in the Purari Delta of Papua New Guinea', in R. Silverman (ed.), *Museum as Process: Translating Local and Global Knowledges*. London: Routledge.

Bendell, J. 2005. 'In Whose Name? The Accountability of Corporate Social Responsibility', *Development in Practice* 15(3–4): 362–74.
Bialecki, J. and G. Daswani. 2015. 'What Is an Individual? The View from Christianity', *HAU: Journal of Ethnographic Theory* 5(1): 271–94.
Bialecki, J., N. Haynes and J. Robbins. 2008. 'The Anthropology of Christianity', *Religion Compass* 2(6): 1139–58.
Blair, M. 1998. 'For Whom Should Corporations be Run? An Economic Rationale for Stakeholder Management', *Long Range Planning* 31(2): 195–200.
———. 2010. 'For Whom Should Corporations Be Run?', in R.A. Phillips and R.E. Freeman (eds), *Stakeholders*. Cheltenham: Edward Elgar.
Blowfield, M. 2007. 'Reasons to Be Cheerful? What We Know about CSR's Impact', *Third World Quarterly* 28(4): 683–95.
———. 2012. 'Business and Development: Making Sense of Business as a Development Agent', *Corporate Governance: International Journal of Business in Society* 12(4): 414–26.
Brumbaugh, R. 1980. 'Models of Separation and a Mountain Ok Religion', *Ethos* 8(4): 332–48.
———. 1990. 'Afek Sang: The Old Woman's Legacy to the Mountain Ok', in B. Craig and D. Hyndman (eds), *Children of Afek: Tradition and Change Among the Mountain-Ok of Central New Guinea*. Oceania Monographs, No. 40. Sydney: University of Sydney.
Brutti, L. 2000. 'Afek's Last Son: Integrating Change in a Papua New Guinea Cosmology', *Ethnohistory* 47(1): 101–11.
———. 2001. 'Where Anthropologists Fear to Tread: Notes and Queries on Anthropology and Consultancy, Inspired by a Fieldwork Experience', *Social Analysis* 45(2): 94–107.
———. 2007. 'From Fertility Rituals to Mining Companies: Eco-Cultural Issues and Land Rights in Oksapmin', *Journal de la Société des Océanistes* 125: 249–55.
Burridge, K. 1960. *Mambu: A Melanesian Millennium*. London: Methuen & Co.
Burton, J. 1997. 'The Principles of Compensation in the Mining Industry', in S. Toft (ed.), *Compensation for Resource Development in Papua New Guinea*. Port Moresby: Law Reform Commission of Papua New Guinea; Canberra: The Australian National University
———. 2007. 'Determinacy of Groups and the "Owned Commons" in Papua New Guinea and Torres Strait"', in J. Weiner and K. Glaskin (eds), *Customary Land Tenure and Registration in Australia and Papua New Guinea: Anthropological Perspectives*. Canberra: ANU E-Press.
Carroll, A. 1999. 'Corporate Social Responsibility: Evolution of a Definitional Construct', *Business Society* 38(3): 268–95.
Casey, E. 1996. 'How to Get from Space to Place in a Fairly Short Stretch of Time', in S. Feld and K. Basso (eds), *Senses of Place*. Santa Fe, NM: School of American Research.

Coffey (Coffey Services Australia Pty Ltd). 2009. 'Environmental Inception Report. Xstrata Frieda River Limited. Frieda River Project'.

———. 2018a. 'Frieda River Limited. Sepik Development Project. Environmental Impact Statement. Appendix 13: Social Impact Assessment'. Last accessed 15 November 2019 from https://friedariver.com/eis/?v=1.

———. 2018b. 'Frieda River Limited. Sepik Development Project. Environmental Impact Statement. Appendix 13: Social Impact Assessment. Appendix 1: Study Area Social Profiles'. Last accessed 15 November 2019 from https://friedariver.com/eis/?v=1.

Collier, P. 2008. 'Laws and Codes for the Resource Curse', *Yale Human Rights and Development Law Journal* 11: 9–281.

Connell, J. 1991. 'Compensation and Conflict: The Bougainville Copper Mine, PNG', in J. Connell and R. Howitt (eds), *Mining and Indigenous Peoples in Australasia*. Sydney: Sydney University Press.

———. 2005. *Papua New Guinea: The Struggle for Development*. Routledge Studies in the Growth Economics of Asia. London: Routledge.

Connell, J. and R. Howitt. 1991. 'Mining, Dispossession, and Development', in J. Connell and R. Howitt (eds), *Mining and Indigenous Peoples in Australasia*. Sydney: Sydney University Press.

Conrad, R. and W. Dye. 1975. 'Some Language Relationships in the Upper Sepik Region of Papua New Guinea', in R. Conrad, W. Dye, N. Thomson and L. Bruce Jr (eds), *Papers in New Guinea Linguistics No. 18*. Canberra: The Australian National University.

Constitution of the Independent State of Papua New Guinea. 1975. Last accessed 15 May 2015 from http://www.paclii.org/pg/legis/consol_act/ cotisopng534/.

Cragg, W. and A. Greenbaum. 2002. 'Reasoning about Responsibilities: Mining Company Managers on What Stakeholders are Owed', *Journal of Business Ethics* 39(3): 319–35.

Craig, B. and D. Hyndman. 1990. *Children of Afek: Tradition and Change Among the Mountain-Ok of Central New Guinea*. Oceania Monographs, No. 40. Sydney: University of Sydney.

Crook, T. 1999. 'Growing Knowledge in Bolivip, Papua New Guinea', *Oceania* 69(4): 225–42.

———. 2004a. 'Combining Rationales from Bolivip: The Person and Property Rights Legislation in Papua New Guinea', in L. Kalinoe and J. Leach (eds), *Rationales of Ownership: Transactions and Claims to Ownership in Contemporary Papua New Guinea*. Wantage: Sean Kingston Publishing.

———. 2004b. 'Transactions in Perpetual Motion', in E. Hirsch and M. Strathern (eds), *Transactions and Creations: Property Debates and the Stimulus of Melanesia*. Oxford: Berghahn.

———. 2007a. '"If You Don't Believe Our Story, At Least Give Us Half of the Money": Claiming Ownership of the Ok Tedi Mine, PNG', *Journal de la Société des Océanistes* 125: 221–28.

———. 2007b. *Anthropological Knowledge, Secrecy and Bolivip, Papua New Guinea: Exchanging Skin*. British Academy Postdoctoral Fellowship Monograph Series. Oxford: Oxford University Press.

———. 2009. 'Exchanging Skin: Making a Science of the Relation between Bolivip and Barth', *Social Analysis* 53(2): 94–107.

Dahl, R. 1961. *Who Governs? Democracy and Power in an American City*. New Haven, CT: Yale University Press.

Demian, M. 2015. 'Dislocating Custom', *PoLAR: Political and Legal Anthropology Review* 38(1): 91–107.

Dening, G. 1989. 'History "in" the Pacific', *The Contemporary Pacific* 1(1–2): 134–39.

———. 1997. 'Empowering Imaginations', *The Contemporary Pacific* 9(2): 419–29.

Dolan, C. and D. Rajak (eds). 2016. *The Anthropology of Corporate Social Responsibility*. New York: Berghahn.

Donaldson, T. and L.E. Preston. 1995. 'The Stakeholder Theory of the Corporation: Concepts, Evidence, and Implications', *Academy of Management Review* 20(1): 65–91.

Douglas, M. 1986. *How Institutions Think*. New York: Syracuse University Press.

Douglas, M. and B. Isherwood. 1979. *The World of Goods: Towards an Anthropology of Consumption*. New York: Basic Books.

Dwyer, P. and M. Minnegal. 2014. 'Where All Rivers Flow West: Maps, Abstraction and Change in the Papua New Guinea Lowlands', *The Australian Journal of Anthropology* 25(1): 37–53.

Englund, H. and J. Leach. 2000. 'Ethnography and the Meta-Narratives of Modernity', *Current Anthropology* 41(2): 225–48.

Ernst, T. 1999. 'Land, Stories, and Resources: Discourse and Entification in Onabasulu Modernity', *American Anthropologist* 101(1): 88–97.

———. 2004. 'Land, Stories and Resources: Some Impacts of Large-Scale Resource Exploitation on Onabasulu Lifeworlds', in A. Rumsay and J. Weiner (eds), *Mining and Indigenous Lifeworlds in Australia and Papua New Guinea*. Wantage: Sean Kingston Publishing.

Errington, F. 1974. 'Indigenous Ideas of Order, Time, and Transition in a New Guinea Cargo Movement', *American Ethnologist* 1(2): 255–67.

Filer, C. 1990 'The Bougainville Rebellion, the Mining Industry and the Process of Social Disintegration in Papua New Guinea', *Canberra Anthropology* 13(1): 1–39.

———. 1996. 'Global Alliances and Local Mediations', *Anthropology Today* 12(5): 26–27.

———. 1997a. 'Compensation, Rent and Power in Papua New Guinea', in S. Toft (ed.), *Compensation for Resource Development in Papua New Guinea*. Port Moresby: Law Reform Commission of Papua New Guinea; Canberra: The Australian National University.

———. 1997b. 'The Melanesian Way of Menacing the Mining Industry', in B. Burt and C. Clerk (eds), *Environment and Development in the Pacific Islands*. Pacific Policy Paper 25. Canberra: The Australian National University.

———. 1999. 'The Dialectics of Negation and Negotiation in the Anthropology of Mineral Resource Development in Papua New Guinea', in A. Cheater (ed.), *The Anthropology of Power: Empowerment and Disempowerment in Changing Structures*. New York: Routledge.

———. 2006. 'Custom, Law and Ideology in Papua New Guinea', *The Asia Pacific Journal of Anthropology* 7(1): 65–84.

———. 2011. 'New Land Grab in Papua New Guinea', *Pacific Studies* 34(2–3): 269–94.

———. 2012. 'Why Green Grabs Don't Work in Papua New Guinea', *Journal of Peasant Studies* 39(2): 599–617.

Filer, C. and M. Macintyre. 2006. 'Grass Roots and Deep Holes: Community Responses to Mining in Melanesia', *The Contemporary Pacific* 18(2): 215–31.

Filer, C., J. Burton and G. Banks. 2008. 'The Fragmentation of Responsibilities in the Melanesian Mining Sector', in C. O'Faircheallaigh and S. Ali (eds), *Earth Matters: Indigenous Peoples, Extractive Industries and Corporate Social Responsibility*. Sheffield: Greenleaf.

Filer, C. and B. Imbun. 2009 'A Short History of Mineral Development Policies in Papua New Guinea, 1972–2002', R.J. May (ed.), *Policy Making and Implementation: Studies from Papua New Guinea*. Canberra: ANU E Press.

Filer, C. and P.Y. Le Meur. 2017. 'Large-Scale Mines and Local-Level Politics', in C. Filer and P.Y. Le Meur (eds), *Large-Scale Mines and Local-Level Politics: Between New Caledonia and Papua New Guinea*. Canberra: ANU Press.

Fletcher, L. and E. Peni. 2019. 'The River Is Not Ours: The Frieda River Mine and the Threat to the Sepik', *Jubilee Australia*. Last accessed 15 September 2019 from https://www.jubileeaustralia.org/project-info/the-frieda-river-mine-the-threat-to-the-sepik-river.

Frederiksen, T. 2018. 'Corporate Social Responsibility, Risk and Development in the Mining Industry', *Resources Policy* 59: 495–505.

Freeman, R.E. 2010. 'Stakeholder Management: Framework and Philosophy', in R.A. Phillips and R.E. Freeman (eds), *Stakeholders*. Cheltenham: Edward Elgar.

Friend, T. 1979. Telefolmin No. 7 of 1979 District Patrol Report (unpublished).

Frynas, J. 2005. 'The False Developmental Promise of Corporate Social Responsibility: Evidence from Multinational Oil Companies', *International Affairs* 81(3): 581–98.

Gardner, D. 1980. 'The Determinants of Mianmin Settlement and Residential Patterns: A Reply to Morren', *Mankind* 12(3): 215–25.

———. 1981. 'Cult Ritual and Social Organisation among the Mianmin', PhD thesis. Australian National University.

———. 1983. 'Performativity in Ritual: The Mianmin Case', *Man*, New Series 18(2): 346–60.

———. 1987. 'Spirits and Conceptions of Agency Among the Mianmin of Papua New Guinea', *Oceania* 57(3): 161–77.

———. 2004. 'Continuity and Identity: Mineral Development, Land Tenure and "Ownership" Among the Northern Mountain Ok', in A. Rumsey and J. Weiner (eds), *Mining and Indigenous Lifeworlds in Australia and Papua New Guinea*. Wantage: Sean Kingston Publishing.

Gell, A. 1999. *The Art of Anthropology: Essays and Diagrams*. London: Athlone Press.

Gellner, E. 1964. *Thought and Change*. London: Weidenfeld and Nicolson.

Gelu, A. 2010. 'Politics and Governance', in T. Webster and L. Duncan (eds), *Papua New Guinea's Development Performance 1975–2008*. The National Research Institute Papua New Guinea Monograph No. 41. Port Moresby: NRI.

Genasci, M. and S. Pray. 2008. 'Extracting Accountability: The Implications of the Resource Curse for CSR Theory and Practice', *Yale Human Rights and Development Law Journal* 11: 37–58.

Gerritsen, R. and M. Macintyre. 1991. 'Dilemmas of Distribution: The Misima Gold Mine, PNG', in J. Connell and R. Howitt (eds), *Mining and Indigenous Peoples in Australasia*. Sydney: Sydney University Press.

Gilberthorpe, E. 2007. 'Fasu Solidarity: A Case Study of Kin Networks, Land Tenure, and Oil Extraction in Kutubu, Papua New Guinea', *American Anthropologist* 109(1): 101–12.

———. 2013a 'In the Shadow of Industry: A Study of Culturization in Papua New Guinea', *Journal of the Royal Anthropological Institute* 19(2): 261–78.

———. 2013b. 'Community Development in Ok Tedi, Papua New Guinea: The Role of Anthropology in the Extractive Industries', *Community Development Journal* 48(3): 466–83.

Gilberthorpe, E. and G. Banks. 2012. 'Development on Whose Terms? CSR Discourse and Social Realities in Papua New Guinea's Extractive Industries Sector', *Resources Policy* 37(2): 185–93.

Gilberthorpe, E. and P. Sillitoe. 2009. 'A Failure of Social Capital: Lessons from Papua New Guinea in the Current Economic Crisis', *Anthropology News* 50(7): 15–16.

Gillespie, K. 2013. 'Ethnomusicology and the Mining Industry: A Case Study from Lihir, Papua New Guinea', *Musicology Australia* 35(2): 178–90.

Goddard, M. 2013. 'Knowing and the Truth: Three Histories of Daugo Island, Papua New Guinea', *The Australian Journal of Anthropology* 24(2): 131–47.

Golub, A. 2006. 'Who Is the "Original Affluent Society"? Ipli "Predatory Expansion" and the Porgera Gold Mine, Papua New Guinea', *The Contemporary Pacific* 18(2): 265–92.

———. 2007. 'From Agency to Agents: Forging Landowner Identities in Porgera', in J. Weiner and K. Glaskin (eds), *Customary Land Tenure and Registration in Australia and Papua New Guinea: Anthropological Perspectives*. Canberra: ANU Press.

———. 2014. *Leviathans at the Gold Mine: Creating Indigenous and Corporate Actors in Papua New Guinea*. Durham, NC: Duke University Press.

Goodstein, J.D. and A.C. Wicks. 2007. 'Corporate and Stakeholder Responsibility: Making Business Ethics a Two-Way Conversation', *Business Ethics Quarterly* 17(3): 375–98.

Gray, R., R. Kouhy and S. Lavers. 1995. 'Corporate Social and Environmental Reporting: A Review of the Literature and a Longitudinal Study of UK Disclosure', *Accounting, Auditing & Accountability Journal* 8(2): 47–77.

Gray, R., Owen, D. and C. Adams. 1996. *Accounting and Accountability: Changes and Challenges in Corporate Social and Environmental Reporting*. London: Prentice Hall.

Green, M. and J. Crouch. 2018. 'Frieda River Limited. Sepik Development Project. Environmental Impact Statement. Appendix 13: Social Impact Assessment. Appendix 2. Cultural Heritage: Baseline and Impact Assessment'. Last accessed 15 November 2018 from https://friedariver.com/eis/?v=1.

Guddemi, P. 1997. 'Continuities, Contexts, Complexities, and Transformations: Local Land Concepts of a Sepik People Affected by Mining Exploration', *Anthropological Forum* 7(4): 629–48.

Hacking, I. 1986. 'Making Up People', in T. Heller, M. Sosna and D. Wellbery (eds), *Reconstructing Individualism: Autonomy, Individuality and the Self in Western Thoughts*. Stanford, CA: Stanford University Press.

———. 1995a. 'The Looping Effects of Human Kinds', in D. Sperber, D. Premack and A.J. Premack (eds), *Casual Cognition: A Multidisciplinary Debate*. Oxford: Clarendon Press.

———. 1995b. *Rewriting the Soul: Multiple Personality and the Sciences of Memory*. Princeton, NJ: Princeton University Press.

Hann, C.M. (ed.). 1998. *Property Relations: Renewing the Anthropological Tradition*. Cambridge: Cambridge University Press.

Harrison, S. 1982. 'Yams and the Symbolic Representation of Time in a Sepik River Village', *Oceania* 53(2): 141–62.

———. 1995. 'Anthropological Perspectives: On the Management of Knowledge', *Anthropology Today* 11(5): 10–14.

———. 1999. 'Cultural Boundaries', *Anthropology Today* 15(5): 10–13.

———. 2001. 'The Past Altered by the Present: A Melanesian Village after Twenty Years', *Anthropology Today* 17(5): 3–9.

Hilson, G. 2001. 'Putting Theory into Practice: How Has the Gold Mining Industry Interpreted the Concept of Sustainable Development?', *Mineral Resources Engineering* 10(4): 397–413.

Hinojosa, A.S. et al. 2017. 'A Review of Cognitive Dissonance Theory in Management Research: Opportunities for Further Development', *Journal of Management* 43(1): 170–99.

Hirsch, E. 1990. 'From Bones to Betelnuts: Processes of Ritual Transformation and the Development of National Culture in Papua New Guinea', *Man New Series* 25(1): 18–34.

———. 1999. 'Colonial Units and Ritual Units: Historical Transformation of Persons and Horizons in Highland Papua New Guinea', *Comparative Studies in Society and History* 41(4): 805–28.

———. 2001a. 'Making up People in Papua', *The Journal of the Royal Anthropological Institute* 7(2): 241–56.

———. 2001b. 'When Was Modernity in Melanesia?', *Social Anthropology* 9(2): 131–46.

———. 2001c. 'New Boundaries of Influence in Highland Papua: "Culture", Mining and Ritual Conversions', *Oceania* 71(4): 298–312.

———. 2002. 'Making up People in Papua: Reply to Stewart and Strathern', *The Journal of the Royal Anthropological Institute* 8(2): 369–71.

———. 2004. 'Mining Boundaries and Local Land Narratives (*tidibe*) in the Udabe Valley, Papua New Guinea', in L. Kalinoe and J. Leach (eds), *Rationales of Ownership: Transactions and Claims to Ownership in Contemporary Papua New Guinea*. Wantage: Sean Kingston Publishing.

———. 2006. 'Landscape, Myth and Time', *Journal of Material Culture* 11(1/2): 151–65.

Hunoldt, M., S. Oertel and A. Galander. 2018. 'Being Responsible: How Managers Aim to Implement Corporate Social Responsibility', *Business & Society* https://doi.org/10.1177/0007650318777738.

Hyndman, D. 1987. 'Mining, Modernization, and Moments of Social Protest in Papua New Guinea', *Social Analysis: The International Journal of Anthropology* 21: 20–38.

———. 2001. 'Academic Responsibilities and Representation of the Ok Tedi Crisis in Postcolonial Papua New Guinea', *The Contemporary Pacific* 13(1): 33–54.

Imbun, B.Y. 2007. 'Cannot Manage without the "Significant Other": Mining, Corporate Social Responsibility and Local Communities in Papua New Guinea', *Journal of Business Ethics* 73(2): 177–92.

Ingold, T. 2016. *Evolution of Social Life*. London: Routledge.

Jacka, J. 2002. 'Cults and Christianity among the Enga and Ipili', *Oceania* 72(3): 196–214.

———. 2005. 'Emplacement and Millennial Expectations in an Era of Development and Globalization: Heaven and the Appeal of Christianity for the Ipili', *American Anthropologist* 107(4): 643–53.

———. 2010. 'The Spirits of Conservation: Ecology, Christianity, and Resource Management in Highlands Papua New Guinea', *Journal for the Study of Religion, Nature and Culture* 4(1): 24–47.

———. 2015. *Alchemy in the Rainforest: Politics, Ecology, and Resilience in a New Guinea Mining Area*. Durham, NC: Duke University Press.

Jackson, M. 2002. *The Politics of Storytelling: Violence, Transgression, and Intersubjectivity*. Copenhagen: Museum Tusculanum Press.

Jackson, R. 1991. 'Not Without Influence: Villages, Mining Companies and Government in PNG', in J. Connell and R. Howitt (eds), *Mining and Indigenous Peoples in Australasia*. Sydney: Sydney University Press.

———. 1997. 'Cheques and Balances: Compensation and Mining in Papua New Guinea', in S. Toft (ed.), *Compensation for Resource Development in Papua New Guinea*. Port Moresby: Law Reform Commission of Papua New Guinea; Canberra: The Australian National University.

Jenkins, H. 2004. 'Corporate Social Responsibility and the Mining Industry: Conflicts and Constructs', *Corporate Social Responsibility and Environmental Management* 11(1): 23–34.

Jenkins, R. 2005. 'Globalization, Corporate Social Responsibility and Poverty', *International Affairs* 81(3): 525–40.

Jones, B. 1980. *Consuming Society: Food and Illness among the Faiwol*. PhD thesis. University of Virginia.

Jorgensen, D. 1981. 'Taro and Arrows: Order, Entropy, and Religion among the Telefolmin', PhD thesis. University of British Columbia.

———. 1990a. 'Secrecy's Turn', *Canberra Anthropology* 13(1): 40–47.

———. 1990b. 'Placing the Past and Moving the Present: Myth and Contemporary History in Telefolmin', *Culture* 10(2): 47–56.

———. 1997. 'Who and What Is a Landowner? Mythology and Marking the Ground in a Papua New Guinea Mining Project', *Anthropological Forum* 7(4): 599–627.

———. 2005. 'Third Wave Evangelism and the Politics of the Global in Papua New Guinea: Spiritual Warfare and the Recreation of Place in Telefolmin', *Oceania* 75(4): 444–61.

———. 2006. 'Hinterland History: The Ok Tedi Mine and Its Cultural Consequences in Telefolmin', *The Contemporary Pacific* 18(2): 233–63.

———. 2007a. 'Clan-Finding, Clan-Making and the Politics of Identity at a Papua New Guinea Mining Project', in J. Weiner and K. Glaskin (eds), *Customary Land Tenure and Registration in Australia and Papua New Guinea: Anthropological Perspectives*. Canberra: Australia National University.

———. 2007b. 'Changing Minds: Hysteria and the History of Spirit Mediumship in Telefolmin', in J. Baker (ed.), *The Anthropology of Morality in Melanesia and Beyond*. Aldershot: Ashgate.

———. 2014. 'Mining Narratives and Multiple Geographies in Papua New Guinea: Ok Tedi, the Emerald Cave and Lost Tribes', *Journal de la Société des Océanistes* 138: 23–36.

Kalinoe, L. 2004. 'The Bases of Ownership Claims Over Natural Resources by Indigenous Peoples in Papua New Guinea', in L. Kalinoe and J. Leach (eds), *Rationales of Ownership: Transactions and Claims to Ownership in Contemporary Papua New Guinea*. Wantage: Sean Kingston Publishing.

Keesing, R. 1982. 'Kastom in Melanesia: An Overview', *The Australian Journal of Anthropology* 13(4): 297–301.

———. 1987. 'Anthropology as Interpretative Quest', *Current Anthropology* 28(2): 161–69.

Kemp, D. 2010. 'Community Relations in the Global Mining Industry: Exploring the Internal Dimensions of Externally Oriented Work', *Corporate Social Responsibility and Environmental Management* 17(1): 1–14.

Kemp, D. and J.R. Owen. 2013. 'Community Relations and Mining: Core to Business but Not "Core Business"', *Resources Policy* 38(4): 523–31.

———. 2015. 'Social Science and the Mining Sector: Contemporary Roles and Dilemmas for Engagement', in S. Price and K. Robinson (eds), *Making a Difference? Social Assessment Policy and Praxis and its Emergence in China*. Oxford: Berghahn.

Kempf, W. 1992. '"The Second Coming of the Lord": Early Christianization, Episodic Time, and the Cultural Construction of Continuity in Sibog', *Oceania* 63(1): 72–86.

King, D. 1997. 'The Big Polluter and the Constructing of Ok Tedi: Eco-Imperialism and Underdevelopment along the Ok Tedi and Fly Rivers of Papua New Guinea', in G. Banks and C. Ballard (eds), *The Ok Tedi Settlement: Issues, Outcomes, and Implications*. National Centre for Development Studies, Research School of Pacific and Asian Studies, Canberra: Australia National University.

King, P., W. Lee and V. Warakai. 1985. *From Rhetoric to Reality? Papua New Guinea's Eight Point Plan and National Goals After a Decade*. Port Moresby: University of Papua New Guinea Press.

Kirsch, S. 1996. 'Anthropologists and the Global Alliances', *Anthropology Today* 12(4): 14–16.

———. 1997. 'Indigenous Response to Environmental Impact Along the Ok Tedi', in S. Toft (ed.), *Compensation for Resource Development in Papua New Guinea*. Port Moresby: Law Reform Commission of Papua New Guinea; Canberra: The Australian National University.

———. 2001. 'Property Effects: Social Networks and Compensation Claims in Melanesia', *Social Anthropology* 9(2): 147–63.

———. 2002. 'Anthropology and Advocacy: A Case Study of the Campaign Against the Ok Tedi Mine', *Critique of Anthropology* 22(2): 175–200.

———. 2004. 'Keeping the Network in View: Compensation Claims, Property and Social Relations in Melanesia', in L. Kalinoe and J. Leach (eds), *Rationales of Ownership: Transactions and Claims to Ownership in Contemporary Papua New Guinea*. Wantage: Sean Kingston Publishing.

———. 2006. *Reverse Anthropology: Indigenous Analysis of Social and Environmental Relations in New Guinea*. Stanford: Stanford University Press.

———. 2007. 'Indigenous Movements and the Risks of Counterglobalization: Tracking the Campaign Against Papua New Guinea's Ok Tedi Mine', *American Ethnologist* 34(2): 303–21.

———. 2014a. *Mining Capitalism: The Relationship Between Corporations and Their Critics*. Oakland, CA: University of California Press.

———. 2014b. 'Imagining Corporate Personhood', PoLAR: *Political and Legal Anthropology Review* 37(2): 207–17.

Kirsch, T. 2007. 'An Appeal for Principled Symmetry: Anthropologies in South Africa and Elsewhere', *Anthropology Southern Africa* 30 (3–4): 97–100.

Knauft, B.M. 2010. 'Beyond Polarization and Partition in Melanesian Anthropology: A Comment on Mosko's Partible Penitents', *The Journal of the Royal Anthropological Institute* 16(2): 244–46.

Lattas, A. 1998. *Cultures of Secrecy: Reinventing Race in Bush Kaliai Cargo Cults*. Madison, WI: The University of Wisconsin Press.

———. 2005. 'Capitalizing on Complicity: Cargo Cults and the Spirit of Modernity on Bali Island (West New Britain)', *Ethnohistory* 52(1): 47–80.

Lawrence, P. 1964. *Road Belong Cargo: A Study of the Cargo Movement in the Southern Madang District New Guinea*. Manchester: Manchester University Press.

Leach, J. 2003. *Creative Land: Place and Procreation on the Rai Coast of Papua New Guinea*. Oxford: Berghahn Books.

———. 2004. 'Land, Trees and History: Disputes Involving Boundaries and Identities in the Context of Development', in L. Kalinoe and J. Leach (eds), *Rationales of Ownership: Transactions and Claims to Ownership in Contemporary Papua New Guinea*. Wantage: Sean Kingston Publishing.

———. 2012a. 'Leaving the Magic Out: Knowledge and Effect in Different Places', *Anthropological Forum* 22(3): 251–70.

———. 2012b. 'Step Inside: Knowledge Freely Available', in F.D. Rubio and P. Baert (eds), *The Politics of Knowledge*. London: Routledge.

———. 2014. '"The Time of Money": Property and Sovereignty as Alternative Narratives of Land and Value near the Ramu NiCo Mining Project (Madang, PNG)', *Journal de la Société des Océanistes* 138–139(9): 53–62.

Lélé, S.M. 1991. 'Sustainable Development: A Critical Review', *World Development* 19(6): 607–21.

Lindstrom, L. 1984. 'Doctor, Lawyer, Wise Man, Priest: Big-Men and Knowledge in Melanesia', *Man*, New Series 19(2): 291–309.

———. 1990. *Knowledge and Power in a South Pacific Society*. Washington, DC: Smithsonian Institute Press.

———. 1993a. 'Cargo Cult Culture: Toward a Genealogy of Melanesian Kastom', *Anthropological Forum* 6(4): 495–513.

———. 1993b. *Cargo Cult: Strange Stories of Desire from Melanesia and Beyond*. Honolulu: University of Hawaii Press.

———. 2008. 'Melanesian Kastom and Its Transformations', *Anthropological Forum* 18(2): 161–78.

LiPuma, E. 1998. 'Modernity and Forms of Personhood in Melanesia', in M. Lambek and A. Strathern (eds), *Bodies and Persons: Comparative Perspectives from Africa and Melanesia*. Cambridge: Cambridge University Press.

Lohmann, R. 2000. 'Cultural Reception in the Contact and Conversion History of the Asabano of Papua New Guinea', PhD thesis. University of Wisconsin-Madison.

———. 2001. 'Introduced Writing and Christianity: Differential Access to Religious Knowledge among the Asabano', *Ethnology* 40(2): 93–111.

———. 2003. 'The Supernatural Is Everywhere: Defining Qualities of Religion in Melanesia and Beyond', *Anthropological Forum* 13(2): 175–85.

———. 2007. 'Morals and Missionary Positionality: Diyos of Duranmin', in J. Baker (ed.), *The Anthropology of Morality in Melanesia and Beyond*. Aldershot: Ashgate.

MacDonald, F. 2013. 'Christianity and Culture Change among the Oksapmin of Papua New Guinea', PhD Thesis. Australian National University.

———. 2014. '"Always Been Christian": Mythic Conflation among the Oksapmin of Papua New Guinea', *Anthropological Forum* 24(2): 175–96.

———. 2017. 'Cosmology and Crisis in Oksapmin, Papua New Guinea', *Oceania* 87(2): 156–72.

Macintyre, M. 1995. 'Violent Bodies and Vicious Exchanges: Personification and Objectification in the Massim', *Social Analysis: The International Journal of Social and Cultural Practice* 37: 29–43.

Macintyre, M. and S. Foale. 2013. 'Science, Traditional Ecological Knowledge, and Anthropology: Managing the Impacts of Mining in Papua New Guinea', *Collaborative Anthropologies* 6(1): 399–418.

MacKenzie, M. 1991. *Androgynous Objects: String Bags and Gender in Central New Guinea*. Reading: Harwood Academic Publishers.

May, R. 1998. 'From Promise to Crisis: A Political Economy of Papua New Guinea', in P. Larmour (ed.), *Governance and Reform in the South Pacific*. National Centre for Development Studies (Pacific Policy Paper 23). Canberra:. Australian National University.

May, R. and M. Spriggs (eds). 1990. *The Bougainville Crisis*. Bathurst, NSW: Crawford House Press.

McCormack, F. and K. Barclay (eds). 2013. 'Engaging with Capitalism: Cases from Oceania'. Bingley: Emerald Group.

McDowell, N. 1985. 'Past and Future: The Nature of Episodic Time in Bun', in D. Gewertz and E. Schieffelin (eds), *History and Ethnohistory in Papua New Guinea*. Oceania Monograph No. 28, Sydney: University of Sydney.

———. 1988. 'A Note on Cargo Cults and Cultural Constructions of Change', *Pacific Studies* 11(2): 121–34.

McGavin, K. 2016 'Where Do You Belong? Identity, New Guinea Islanders, and the Power of Peles', *Oceania* 86(1): 57–74.

McVea, J.F. and R.E. Freeman. 2005. 'A Names-and-Faces Approach to Stakeholder Management: How Focusing on Stakeholders as Individuals Can Bring Ethics and Entrepreneurial Strategy Together', *Journal of Management Inquiry* 14(1): 57–69.

Minnegal, M. and P.D. Dwyer. 2017. *Navigating the Future: An Ethnography of Change in Papua New Guinea*. Canberra: ANU Press.

Minnegal, M., S. Lefort and P. Dwyer. 2015. 'Reshaping the Social: A Comparison of Fasu and Kubo-Febi Approaches to Incorporating Land Groups', *The Asia Pacific Journal of Anthropology* 16(5): 496–513.

Moretti, D. 2006. 'The Gender of the Gold: An Ethnographic and Historical Account of Women's Involvement in Artisanal and Small-Scale Mining in Mount Kaindi, Papua New Guinea', *Oceania* 76(2): 133–49.

Morren, G.E. 1979. 'Seasonality among the Miyanmin: Wild Pigs, Movement, and Dual Kinship Organization', *Mankind* 12(1): 1–12.

———. 1981. 'A Small Footnote to the "Big Walk": Environment and Change among the Miyanmin of Papua New Guinea', *Oceania* 52(1): 39–65.

———. 1986. *The Miyanmin: Human Ecology of a Papua New Guinea Society*. UMI Studies in Cultural Anthropology 9. Ann Arbor: UMI Research Press.

Mosko, M. 2010a. 'Partible Penitents: Dividual Personhood and Christian Practice in Melanesia and the West', *Journal of the Royal Anthropological Institute* 16(2): 215–40.

———. 2010b. 'Partible Penitents: A Response to Comments', *Journal of the Royal Anthropological Institute* 16(2): 253–59.

———. 2013. 'Dividuals, Individuals, or Possessive Individuals? Recent Transformations of North Mekeo Commoditization, Personhood, and Sociality', in F. McCormack and K. Barclay (eds), *Engaging with Capitalism: Cases from Oceania*. Bingley: Emerald Group.

———. 2015. 'Unbecoming Individuals: The Partible Character of the Christian Person', *HAU: Journal of Ethnographic Theory* 5(1): 361–93.

Moutu, A. 2008. 'Frieda- Another Ecological Disaster?' (letter), *The National*, 5 May 2008.

MRA (Mineral Resources Authority). 'Mineral Project Life Cycle: Licensing Phases'. Last accessed 15 January 2018 from http://www.mra.gov.pg/Portals/2//Mining%20Info/licensing%20phase.pdf.

Moylan, T. 1979. 'Identification of the Land Owners in the Frieda Area and Their Affiliations With Other Groups' (unpublished).

Newell, P. and J. Frynas. 2007. 'Beyond CSR? Business, Poverty and Social Justice: An Introduction', *Third World Quarterly* 28(4): 669–81.

O'Faircheallaigh, C. 2008. 'Introduction', in C. O'Faircheallaigh and S. Ali (eds), *Earth Matters: Indigenous Peoples, Extractive Industries and Corporate Social Responsibility*. Sheffield: Greenleaf Publishing.

Otto, T. 2015. 'Transformations of Cultural Heritage in Melanesia: From Kastam to Kalsa', *International Journal of Heritage Studies* 21(2): 117–32.

Owen, J.R. 2016. 'Social License and the Fear of Mineras Interruptus', *Geoforum* 77: 102–5.

Owen, J.R. and D. Kemp. 2013. 'Social Licence and Mining: A Critical Perspective', *Resources Policy* 38(1): 29–35.

———. 2014. 'Mining and Community Relations: Mapping the Internal Dimensions of Practice', *The Extractive Industries and Society* 1(1): 12–19.

———. 2017. *Extractive Relations: Countervailing Power and the Global Mining Industry*. London: Routledge.

PanAust. 2018. 'A Nation-Building Development Pathway for the Frieda River Project: Company Announcement'. Last accessed 15 June 2019 from https://panaust.com.au/company-announcements.

Papua New Guinea Environmental Act. 2000. Last accessed 15 April 2015 from http://www.paclii.org/pg/legis/consol_act/ea2000159/.

Papua New Guinea Land Act. 1996. Last accessed 15 April 2015 from http://www.paclii.org/pg/legis/consol_act/la199648/.

Papua New Guinea Mining (Safety) Act. 1977. Last accessed 15 April 2015 from http://www.paclii.org/pg/legis/consol_act/ma1977151/.

Papua New Guinea Mining Act. 1992. Last accessed 15 April 2015 from http://www.paclii.org/pg/legis/consol_act/ma199281/.

Parsons, R., J. Lacey and K. Moffat. 2014. 'Maintaining Legitimacy of a Contested Practice: How the Minerals Industry Understands Its "Social Licence to Operate"', *Resources Policy* 41: 83–90.

Phillips, R. 1997. 'Stakeholder Theory and A Principle of Fairness', *Business Ethics Quarterly* 7(1): 51–66.

Phillips, R., R.E. Freeman and A.C. Wicks. 2010. 'What Stakeholder Theory Is Not', in R. Phillips and R.E. Freeman (eds), *Stakeholders*. Cheltenham: Edward Elgar.

Pijpers, R.J. and T.H. Eriksen. 2018. *Mining Encounters: Extractive Industries in an Overheated World*. London: Pluto Press.

Polier, N. 1992. 'The Mines of Min: History, Gender, and Social Transformation Among the Faiwolmin of Papua New Guinea', PhD Thesis. New School for Social Research.

Power, T. 1997. 'Landowner Compensation: Policy and Practice', in S. Toft (ed.), *Compensation for Resource Development in Papua New Guinea*. Port Moresby: Law Reform Commission of Papua New Guinea; Canberra: The Australian National University.

———. 2008. 'Incorporated Land Groups in Papua New Guinea' in *Making Land Work. Volume 2: Case Studies in Customary Land and Development in the Pacific* Canberra: AusAid. Last accessed 15 September 2017 from http://dfat.gov.au/about-us/publications/Documents/MLW_VolumeTwo_CaseStudy_1.pdf.

PSGTAD (Paiyamo Sane Grup of Telefomin and Ambunti Districts). 2011. 'Paiyamo Sane: Clans, Histories and Their Make Up' (unpublished).

Rajak, D. 2011. *In Good Company: An Anatomy of Corporate Social Responsibility*. Stanford, CA: Stanford University Press.

Ramirez, R. 1999. 'Stakeholder Analysis and Conflict Management', in D. Buckles (ed.), *Cultivating Peace: Conflict and Collaboration in Natural Resource Management*. Ottawa: International Development Research Centre.

Reed, A.M. and D. Reed. 2009. 'Partnerships for Development: Four Models of Business Involvement', *Journal of Business Ethics* 90(1): 3–37.

Reed, M.S. et al. 2009. 'Who's in and Why? A Typology of Stakeholder Analysis Method for Natural Resource Management', *Journal of Environmental Management* 90(5): 1933–49.

Reilly, B. 2008. 'Ethnic Conflict in Papua New Guinea', *Asia Pacific Viewpoint* 49(1): 12–22.

Rivers, J. 1999. 'Formulating Basic Policy for Community Relations Programs', *State, Society and Governance in Melanesia Discussion Paper* No. 99/1. Canberra: Australian National University.

Robbins, J. 1995. 'Dispossessing the Spirits: Christian Transformation of Desire and Ecology among the Urapmin of Papua New Guinea', *Ethnology* 34(3): 211–24.

———. 1998a. 'Becoming Sinners: Christianity and Desire among the Urapmin of Papua New Guinea', *Ethnology* 37(4): 299–316.

———. 1998b. 'On Reading "World News": Apocalyptic Narrative, Negative Nationalism and Transnational Christianity in a Papua New Guinea Society', *Social Analysis* 42(2): 103–30.

———. 2001. 'God Is Nothing but Talk: Modernity, Language and Prayer in a Papua New Guinean Society', *American Anthropologist* 103(4): 901–12.

———. 2007. 'Continuity Thinking and the Problem of Christian Culture Belief: Time, and the Anthropology of Christianity', *Current Anthropology* 48(1): 5–38.

———. 2010. 'Melanesia, Christianity, and Cultural Change: A Comment on Mosko's "Partible Penitents"', *The Journal of the Royal Anthropological Institute* 16(2): 241–43.

Robbins, J. and H. Wardlow (eds). 2005. *The Making of Global Modernities in Melanesia: Humiliation, Transformation and the Nature of Cultural Change*. Aldershot: Ashgate.

Rumsey, A. and J. Weiner (eds). 2001. *Emplaced Myth: Space, Narrative and Knowledge in Aboriginal Australia and Papua New Guinea*. Honolulu: University of Hawaii Press.

———. 2004. *Mining and Indigenous Lifeworlds in Australia and Papua New Guinea*. Wantage: Sean Kingston Publishing.

Sachs, J. and A. Warner. 2001. 'The Curse of Natural Resources', *European Economic Review* 45(4-6): 827–38.

Sagir, B. 2004. 'The Politics of Petroleum Extraction and Royalty Distribution at Lake Kutubu', in A. Rumsey and J. Weiner (eds), *Mining and Indigenous Lifeworlds in Australia and Papua New Guinea*. Wantage: Sean Kingston Publishing.

Sahlins, M. 1983. 'Other Times, Other Customs: The Anthropology of History', *American Anthropologist* 85(3): 517–44.

———. 1985. *Islands of History: Ethnology and History*. Chicago: University of Chicago Press.

———. 1992. 'The Economics of Develop-Man in the Pacific', *Res: Anthropology and Aesthetics* 21(1): 12–25.

———. 1999. 'Two or Three Things That I Know about Culture', *The Journal of the Royal Anthropological Institute* 5(3): 399–21.

Scaglion, R. 1999. 'Yam Cycles and Timeless Time in Melanesia', *Ethnology* 38(3): 211–25.

———. 2000. 'Afterword: Zero Hour: Reflecting Backward, Looking Forward', *Ethnohistory* 47(1): 227–40.

Schram, R. 2015. 'A Society Divided: Death, Personhood, and Christianity in Auhelawa, Papua New Guinea', *HAU: Journal of Ethnographic Theory* 5(1): 317–37.

Schuster, M. 1990. 'Aspects of the Aibom Concept of History', in N. Lutkehaus et al. (eds), *Sepik Heritage: Tradition and Change in Papua New Guinea*. Durham, NC: Carolina Academic Press.

Scott, J.C. 1998. *Seeing Like a State: How Certain Schemes to Improve Human Condition Have Failed.* New Haven: Yale University Press.

Schorch, P. and A. Pascht. 2017. 'Reimagining Oceania through Critical Junctures—Introduction', *Oceania* 87(2): 114–23.

Sillitoe, P. 1998. 'The Development of Indigenous Knowledge: A New Anthropology', *Current Anthropology* 39(2): 223–52.

———. 1999. 'Beating the Boundaries: Land Tenure and Identity in the Papua New Guinea Highlands', *Journal of Anthropological Research* 55(3): 331–60.

Slotta, J. 2014. 'Revelations of the World: Transnationalism and the Politics of Perceptions in Papua New Guinea', *American Anthropologist* 116(3): 626–42.

Smith, K. 2012. 'From Dividual and Individual Selves to Porous Subjects', *The Australian Journal of Anthropology* 23(1): 50–64.

Stead, V. 2017. *Becoming Landowners: Entanglements of Custom and Modernity in Papua New Guinea and Timor-Leste*. Honolulu: University of Hawai'i Press.

Stewart, C. 2016. 'Historicity and Anthropology', *Annual Review of Anthropology* 45: 79–94.

Stewart, P.J. and A. Strathern. 2000a. *Identity Work: Constructing Pacific Lives.* ASAO Monograph Series. Pittsburgh, PA: University of Pittsburgh Press.

———. (eds). 2000b. 'Millennial Countdown in New Guinea', Special Issue of *Ethnohistory* 47(1).

———. 2002. 'Comments on "Making up People in Papua"', *The Journal of the Royal Anthropological Institute* 8(2): 367–71.

Strathern, A. and P.J. Stewart. 1998a. 'Seeking Personhood: Anthropological Accounts and Local Concepts in Mount Hagen, Papua New Guinea', *Oceania* 68(3): 170–88.

———. 1998b. 'Shifting Places, Contested Spaces: Land and Identity Politics in the Pacific', *The Australian Journal of Anthropology* 9(2): 209–24.

Strathern, M. 1988. *The Gender of the Gift*. Berkeley, CA: University of California Press.

———. 1991. 'One Man and Many Men', in M. Godelier and M. Strathern (eds), *Big*

Men and Great Men: Personifications of Power in Melanesia. Cambridge: Cambridge University Press.
———. 1996. 'Cutting the Network', *Journal of the Royal Anthropological Institute* 2(3): 517–35.
———. 1999. *Property, Substance and Effect: Anthropological Essays on Persons and Things*. London: Athlone Press.
———. 2004. 'Introduction: Rationales of Ownership', in L. Kalinoe and J. Leach (eds), *Rationales of Ownership: Transactions and Claims to Ownership in Contemporary Papua New Guinea*. Wantage: Sean Kingston Publishing.
———. 2005. *Kinship, Law and the Unexpected: Relatives Are Always a Surprise*. Cambridge: Cambridge University Press.
Telban, B. 2013. 'The Power of Place: Spatio-Temporality of a Melanesian Religious Movement', *Anthropological Notebooks* 19(3): 81–100.
———. 2017. 'Seeing and Holding Time: Karawari Perceptions of Temporalities, Calendars and Clocks', *Time & Society* 26(2): 182–202.
Tilton, J. 1996. 'Exhaustible Resources and Sustainable Development: Two Different Paradigms', *Resources Policy* 22(1–2): 91–97.
Toft, S. 1997. 'Patrons or Clients? Aspects of Multinational Capital-Landowner Relations in Papua New Guinea', in S. Toft (ed.), *Compensation for Resource Development in Papua New Guinea*. Port Moresby: Law Reform Commission of Papua New Guinea; Canberra: The Australian National University.
Turner, B.S. 1988. 'Individualism, Capitalism and the Dominant Culture: A Note on the Debate', *The Australian and New Zealand Journal of Sociology* 24(1): 47–64.
Valenta, R.K et al. 2019. 'Re-Thinking Complex Orebodies: Consequences for the Future World Supply of Copper', *Journal of Cleaner Production* 220: 816–26.
Votaw, D. 1973. 'Genius Becomes Rare: A Comment on the Doctrine of Social Responsibility', *California Management Review* 15(3): 5–19.
Wagner, R. 1981. *The Invention of Culture: Revised and Expanded Edition*. Chicago: University of Chicago Press.
———. 1984. 'Ritual as Communication: Order, Meaning, and Secrecy in Melanesian Initiation Rites', *Annual Review of Anthropology* 13(1): 143–55.
———. 1986. *Symbols That Stand for Themselves*. Chicago: University of Chicago Press.
———. 1991. 'The Fractal Person', in M. Godelier and M. Strathern (eds), *Big Men and Great Men*. Cambridge: Cambridge University Press.
Wardlow, H. 2004. 'The Mount Kare Python: Huli Myths and Gendered Fantasies of Agency', in A. Rumsey and J. Weiner (eds), *Mining and Indigenous Lifeworlds in Australia and Papua New Guinea*. Wantage: Sean Kingston Publishing.
Wassmann, J. 1990. 'The Nyuara Concepts of Space and Time', N. Lutkehaus et al. (eds), *Sepik Heritage: Tradition and Change in Papua New Guinea*. Durham, NC: Carolina Academic Press.

WCED (World Commission on Environment and Development). 1987. 'Our Common Future'. Oxford, Oxford University Press.

Weiner, J.F. 1991. *The Empty Space: Poetry, Space, and Being among the Foi of Papua New Guinea*. Indianapolis: Indiana University Press.

———. 1995. 'Anthropologists, Historians and the Secret of Social Knowledge', *Anthropology Today* 11(5): 3–7.

———. 1998a. 'The Incorporated Ground: The Contemporary Work of Distribution in the Kutubu Oil Project Area, Papua New Guinea', *Resource Management in Asia Pacific Working Paper no. 17*. Canberra: Australia National University.

———. 1998b. 'Revealing the Grounds of Life in Papua New Guinea', *Social Analysis* 42(3): 135–42.

———. 2001. 'The Foi Incorporated Land Group: Law and Custom in Group Definition and Collective Action in the Kutubu Oil Project Area, PNG', *State, Society and Governance in Melanesia Program Working Paper no. 2*. Canberra: Australia National University.

———. 2002. 'The Work of Inscription in Foi Poetry', in B. David and M. Wilson (eds), *Inscribed Landscapes: Marking and Making Place*. Honolulu: University of Hawaii Press.

———. 2004. 'Introduction: Depositings', in A. Rumsay and J. Weiner (eds), *Mining and Indigenous Lifeworlds in Australia and Papua New Guinea*. Wantage: Sean Kingston Publishing.

———. 2007. 'The Foi Incorporated Land Group: Group Definition and Collective Action in the Kutubu Oil Project, Papua New Guinea', in J. Weiner and K. Glaskin (eds), *Customary Land Tenure and Registration in Australia and Papua New Guinea: Anthropological Perspectives*. Canberra: ANU Press.

Weiner, J.F. and K. Glaskin (eds). 2007. *Customary Land Tenure and Registration in Australia and Papua New Guinea: Anthropological Perspectives*. Canberra: ANU Press.

Welker, M. 2014. *Enacting the Corporation: An American Mining Firm in Post-Authoritarian Indonesia*. Berkeley, CA: University of California Press.

West, P. 2006 'Environmental Conservation and Mining: Strange Bedfellows in the Eastern Highlands of Papua New Guinea', *The Contemporary Pacific* 18(2): 295–313.

Wilson, D. 2008. 'Defining a Sustainable Project' (letter), *The National*, 16 July 2008.

Wolfe, R.A. and D.S. Putler. 2002. 'How Tight Are the Ties that Bind Stakeholder Groups?', *Organization Science* 13(1): 64–80.

Worsley, P. 1968. *The Trumpet Shall Sound: A Study of 'Cargo' Cults in Melanesia*. New York: Shocked Books.

Xstrata Copper. 2008. 'Frieda River Project Sustainability Report 2007'. Brisbane: Xstrata Copper.

———. 2009. 'Frieda River Project Sustainability Report 2008'. Brisbane: Xstrata Copper.

———. 2010. 'Frieda River Project Sustainability Report 2009'. Brisbane: Xstrata Copper.

———. 2011. 'Frieda River Project Sustainability Report 2010'. Brisbane: Xstrata Copper.

———. 2012. 'Frieda River Project Sustainability Report 2011'. Brisbane: Xstrata Copper.

———. 2013. 'Frieda River Project Sustainability Report 2012'. Brisbane: Xstrata Copper.

Index

www.ingramcontent.com/pod-product-compliance
Ingram Content Group UK Ltd.
Pitfield, Milton Keynes, MK11 3LW, UK
UKHW021837150926
8656IPUK00004B/25

9 781789 208566